INTERPRETING GÖDEL: CRITICAL ESSAYS

The logician Kurt Gödel (1906–1978) published a paper in 1931 formulating what have come to be known as his "incompleteness theorems," which prove, among other things, that within any formal system with resources sufficient to code arithmetic, questions exist which are neither provable nor disprovable on the basis of the axioms which define the system. These are among the most celebrated results in logic today. In this volume, leading philosophers and mathematicians assess important aspects of Gödel's work on the foundations and philosophy of mathematics. Their essays explore almost every aspect of Gödel's intellectual legacy including his concepts of intuition and analyticity, the Completeness Theorem, the set-theoretic multiverse, and the state of mathematical logic today. This groundbreaking volume will be invaluable to students, historians, logicians, and philosophers of mathematics who wish to understand the current thinking on these issues.

JULIETTE KENNEDY is an Associate Professor in the Department of Mathematics and Statistics at the University of Helsinki.

INTERPRETING GÖDEL

Critical Essays

EDITED BY

JULIETTE KENNEDY
University of Helsinki

CAMBRIDGE
UNIVERSITY PRESS

CAMBRIDGE
UNIVERSITY PRESS

University Printing House, Cambridge CB2 8BS, United Kingdom

Cambridge University Press is part of the University of Cambridge.

It furthers the University's mission by disseminating knowledge in the pursuit of education, learning and research at the highest international levels of excellence.

www.cambridge.org
Information on this title: www.cambridge.org/9781316639771

© Cambridge University Press 2014

First published 2014
First paperback edition 2016

A catalogue record for this publication is available from the British Library

Library of Congress Cataloguing in Publication data
Interpreting Gödel : critical essays / edited by Juliette Kennedy, University of Helsinki.
pages cm
ISBN 978-1-107-00266-1 (Hardback)
1. Logic, Symbolic and mathematical.   2. Gödel, Kurt.   3. Mathematics–Philosophy.
I. Kennedy, Juliette, 1955– editor of compilation.
QA9.2.I58 2014
511.3–dc23   2014007629

ISBN 978-1-107-00266-1 Hardback
ISBN 978-1-316-63977-1 Paperback

This book is dedicated to the memory of my mother, Poppy Kennedy

Contents

List of contributors   *page* ix
Acknowledgements   xi

1   Introduction. Gödel and analytic philosophy: how did
    we get here?   1
    Juliette Kennedy

PART I   GÖDEL ON INTUITION   9

2   Intuitions of three kinds in Gödel's views on the continuum   11
    John P. Burgess

3   Gödel on how to have your mathematics and know it too   32
    Janet Folina

PART II   THE COMPLETENESS THEOREM   57

4   Completeness and the ends of axiomatization   59
    Michael Detlefsen

5   Logical completeness, form, and content: an archaeology   78
    Curtis Franks

PART III   COMPUTABILITY AND ANALYTICITY   107

6   Gödel's 1946 Princeton bicentennial lecture: an appreciation   109
    Juliette Kennedy

7   Analyticity for realists   131
    Charles Parsons

vii

PART IV   THE SET-THEORETIC MULTIVERSE   151

8   Gödel's program   153
    John R. Steel

9   Multiverse set theory and absolutely undecidable propositions   180
    Jouko Väänänen

PART V   THE LEGACY   209

10 Undecidable problems: a sampler   211
    Bjorn Poonen

11 Reflecting on logical dreams   242
    Saharon Shelah

Bibliography   256
Index   277

Contributors

JOHN BURGESS is the John N. Woodhull Professor of Philosophy and Associated Faculty of the Department of Mathematics at Princeton University. His latest book is entitled *Saul Kripke: Puzzles and Mysteries* (2013).

JANET FOLINA is Professor of Philosophy at Macalester College. Her most recent papers include a survey of nineteenth century philosophy of mathematics entitled "1790–1870: Some developments in the philosophy of mathematics" (2012) and "Hamilton and Newton: in defense of truth in algebra" (2012).

MICHAEL DETLEFSEN is the McMahon-Hank Professor of Philosophy at the University of Notre Dame. His recent papers include "Freedom and consistency" (2013), "Gentzen's anti-formalist ideas" (2013) and "Dedekind against intuition: Rigor, scope and the motives of his logicism" (2011).

CURTIS FRANKS is Associate Professor of Philosophy at the University of Notre Dame. The author of *The Autonomy of Mathematical Knowledge* (2009), his other papers on Gödel include "The Gödelian inferences" (2009) and "Stanley Tennenbaum's Socrates" (2012).

JULIETTE KENNEDY is Associate Professor in the Department of Mathematics and Statistics of the University of Helsinki. Her recent papers on Gödel include "Gödel's thesis: An appreciation," in *Kurt Gödel and the Foundations of Mathematics: Horizons of Truth* (2011).

CHARLES PARSONS is the Edgar Pierce Professor of Philosophy, Emeritus at Harvard University. He was editor, with Solomon Feferman and others, of Volume III of the collected works of Kurt Gödel, *Unpublished Essays and Lectures* (1995) and of Volumes IV and V, *Correspondence* (2003). His most recent book is entitled *Mathematical Thought and its Objects*, (2008).

JOHN STEEL is Professor of Mathematics at the University of California at Berkeley. His most recent papers are "Ordinal definability in models of determinacy" and "HOD as a core model" (the latter joint with W. H. Woodin), both in *Ordinal Definability and Recursion Theory: the New Cabal Seminar*, Volume III (2012).

BJORN POONEN is the Claude Shannon Professor of Mathematics at the Massachusetts Institute of Technology. Among Poonen's major recent works are the paper "Insufficiency of the Brauer–Manin obstruction applied to étale covers" (2010) and "Chabauty's method proves that most odd degree hyperelliptic curves have only one rational point," with Michael Stoll (2013).

SAHARON SHELAH is Professor of Mathematics at the Hebrew University of Jerusalem and Rutgers University. The author of over 1000 mathematical papers and books and the recipient of the Erdős, Bolyai and Wolf prizes, among others, he was most recently awarded the Leroy P. Steele Prize for Lifetime Achievement (Seminal Contribution to Research) for his book, *Classification Theory and the Number of Nonisomorphic Models* (1990).

JOUKO VÄÄNÄNEN is Professor of Mathematics at the University of Helsinki and Professor of Mathematical Logic and Foundations of Mathematics at the University of Amsterdam. His recent publications include the book *Models and Games* (2011), and the paper "Second order logic or set theory?" (2012).

Acknowledgements

I wish to thank Hilary Gaskin of Cambridge University Press for her support of this volume. The volume was completed while I was a member of the School of Historical Studies of the Institute for Advanced Study. I wish to express my gratitude to the Institute and to the School of Historical Studies through its Otto Neugebauer Fund, as well as to the Academy of Finland and the Väisälä Foundation for this support. Finally I wish to thank Marcia Tucker, Librarian of the Historical Studies and Social Science Library of the IAS, and her staff, for all of their cheerful assistance over the years.

CHAPTER I

Introduction. Gödel and analytic philosophy: how did we get here?

Juliette Kennedy

1 Introduction

It is often said about Kurt Gödel that he was the greatest logician of the twentieth century. His work in mathematical logic, when it does not constitute the very ground out of which its various subfields grew and developed, made the continuation of the subject possible at a time when fundamental concepts had not even been identified, and proofs of key theorems – in those cases when they had been stated – had not material-ized in anything like their final form. This is not to say that Gödel was intellectually infallible; one could also point to the richness of Gödel's logical milieu. But there is no doubt that a gigantic intelligence had turned to the field of mathematical logic – and how much better off the subject was for it!

Gödel's philosophical work on the other hand, work to which he devoted himself almost exclusively from the mid-1940s until his death in 1978, has not been as well received. Put another way, any praise of Gödel's contributions to the foundations of mathematics has largely been limited to his theorems.[1] Gödel the *philosopher* – and indeed even today it is a matter of debate, whether Gödel can be regarded as a philosopher at all – has traditionally been seen as advocating a crude form of Platonism in his philosophical writings, one entangled with the views of Kant and Leibniz in a way which was seen as philosophically naive and primarily historical; and one which, anachronistically, seemed to give no quarter to what turned out to be the single most important development in twentieth

---

[1] See for example Boolos's introduction (Gödel 1995, pp. 290–304) to Gödel's posthumously published 1951 Gibbs Lecture ("Some basic theorems of the foundations of mathematics and their philosophical implications," reprinted in Gödel (1995), pp. 304–323):

> What may be found problematic in Gödel's judgement that his conclusion is of philosophical interest is that it is certainly not obvious what it means to say that the human mind... is a Turing machine.

century (analytic) philosophy, namely the so-called linguistic turn inaugurated by Frege, Russell and Moore. To the contrary, Gödel's Platonism, that is to say his various formulations of the view that mathematics is contentual, or in other versions that mathematical truth is bivalent, or in still other versions that mathematical objects enjoy some positive sense of existence, were seen by philosophers – when they did not simply bypass his work – as the antiquarian views of an old-fashioned, albeit great mathematician, untrained in philosophy and nostalgic for the days when the concept of mathematical truth was considered to be beyond criticism – an ironic development in the light of Gödel's actual discoveries.

With this volume we wish to effect a change in the philosophical body politic; to call attention to threads in Gödel's thinking which have turned out to be, in light of the directions in which philosophy has developed since Gödel's time, either newly or persistently important. We wish to reassess Gödel's practice of *philosophy as mathematics*; in a word, to reassess his philosophical work in the light of possibly favorable developments. Recent excursions into mathematical naturalism, for example, to be found in works by Penelope Maddy and others, have brought into the philosophy of mathematics a newly invigorated focus on mathematical practice – a nonnegotiable, core commitment for Gödel. Of course, much of the writing on Gödel's philosophical work has focused on his avowed Platonism. And while there is every reason to expect that Gödel will continue to be a canonical representative of that view in the minds of many philosophers, others have gained philosophical traction in areas of Gödel's writings which are less overtly metaphysical and more oriented toward actual mathematics, set theory in particular, but also other material which is "closer to the ground" mathematically and logically.

Of Gödel's philosophically informed logical work, his Completeness Theorem is a fundamental technical result. But the resurgence of interest in logical consequence places it at the center of contemporary philosophical focus. As Curtis Franks puts it in Chapter 5,

> While the theorem contained in Gödel's thesis is a cornerstone of modern logic, its far more sweeping and significant impact is the fact that, through its position in a network of technical results and applications, the way of thinking underlying the result has come to seem definitive and necessary, to the extent that we have managed to forget that it has not always been with us.

Franks's observation that as far as the concept of logical consequence goes, our world is Gödelian through and through, could equally well apply to other projects within the contemporary philosophy of mathematics enterprise. Gödel's trademark as a philosopher, his *modus operandi* as it were, was to practice philosophy as if it were mathematics; to conjure sharp, mathematical conjectures out of inchoate philosophical material, refashioning that material so as to be subject to proof. It was a radical approach to philosophical practice, harkening back to Leibniz's *calculemus*, if not to the calculating machines of Ramon Lull, as well as to the Husserlian project of *Philosophie als Strenge Wissenschaft*. It is ironic that if one scrutinizes Gödel's philosophical writings in the light of his own standards, this renders much of it ungrounded; and indeed, Gödel often remarked of his writings that short of a more exact treatment, much of what had been laid out there was not to be taken as definitive. On the other hand – and this is the subtlety here – Gödel had a very broad notion of proof.[2]

The method was found uncongenial at the time. For example George Boolos's 1995 introduction to Gödel's posthumously published 1951 Gibbs Lecture, about the philosophical consequences of the Incompleteness Theorems, results which, according to Gödel, "...have not been adequately discussed, or have only just been taken notice of" (Gödel 1995, p. 305), contains the following statement:

> Gödel's idea that we shall one day achieve sufficient clarity about the concepts involved in the philosophical discussion to be able to prove, mathematically, the truth of some position in the philosophy of mathematics, however, appears significantly less credible at present than his Platonism. (Gödel 1995, p. 290)

We referred above to favorable developments. In the years since Boolos wrote his introduction, the field of philosophy of mathematics has begun to shift toward, one might say, the concrete. Episodes in the history of mathematics – odd accidents, moments of perplexity, turning points and the like – whose philosophical significance was previously overlooked, are now given sustained and detailed treatment, "under a microscope" so to speak, and "pushed to the limit."[3] "Conceptual fine-structure" is a term of art; and while the a priorist tradition – championed, of course, by Gödel, but from a standpoint that was centered within the practice – is still very dominant, for a significant percentage of philosophers of mathematics, the importance of a priorism has begun to recede. Franks described this

---

[2] See, for example, Gödel (1946), "Remarks before the Princeton bicentennial conference of problems in mathematics," reprinted in Gödel (1990).

[3] See, for example, Arana and Mancosu (2012), Wilson's massive book (2008) and Brandom (2011), a review of Wilson (2008).

shift away from a priori or *second-order* philosophical discourse (in his terminology) as follows:

> Recent philosophical writing about mathematics has largely abandoned the *a priorist* tradition and its accompanying interest in grounding mathematical activity. The foundational schools of the early twentieth century are now treated more like historical attractions than like viable ways to enrich our understanding of mathematics. This shift in attitudes has resulted not so much from a piecemeal refutation of the various foundational programs, but from the gradual erosion of interest in laying foundations, from our culture's disenchantment with the idea that a philosophical grounding may put mathematical activity in plainer view, make more evident its rationality, or explain its ability to generate a special sort of knowledge about the world. (Franks 2009, p. 169)

Our point is this: the role of the *practitioner*, in philosophy of mathematics but also in other philosophical subfields, has now become central. The consequence for set theory in particular is that this wholly mathematical project is now beginning to be perceived as a wholly philosophical one as well – surely a mark of our Gödelian inheritance, and one we take particular note of in this volume.

Rather than a comprehensive survey, our volume is more of a snapshot of the contemporary take on Gödel's work. We suspend judgment on the taxonomy of subjects, appropriating for *philosophy* issues like the decidability of diophantine equations, the generic multiverse in set theory and Shelah's Main Gap program in model theory. Chapters on these topics by Poonen, Steel, Väänänen and Shelah, respectively, are set side by side and on an equal basis (in terms of philosophical interest) with chapters on intuition by Folina and Burgess, on logical consequence by Detlefsen and Franks, and on analyticity by Parsons.

Practical considerations may undermine comprehensiveness in an editorial volume, and ours is no exception. (For example, Gödel's massive contribution to intuitionism is only touched on in some of the essays here, and only in passing.) We thus take this opportunity to refer the reader to the recent work of a few important commentators whose work does not appear here: in addition to W. W. Tait's work on Gödel and intuitionism, D. A. Martin's papers on Gödel's conceptual realism; P. Maddy's work on perceptual realism and more recently on naturalism in set theory; H. Woodin's work on the multiverse and on the decidability of the continuum problem generally; finally, R. Tieszen and M. van Atten's work on Gödel and phenomenology.

2 Gödel's Platonism: a case study in method

The term Platonism as it is used in the current context of philosophy of mathematics seems to have been coined by Bernays in his 1934 lecture "On Platonism in mathematics":

> ... the objects of a theory are viewed as elements of a totality such that one can reason as follows: For each property expressible using the notions of the theory, it is [an] objectively determinate [fact] whether there is or there is not an element of the totality which possesses this property ... the tendency of which we are speaking consists in viewing the objects as cut off from all links to the reflecting subject. Since this tendency asserted itself especially in the philosophy of Plato, allow me to call it "Platonism." (Reprinted in Benacerraf and Putnam 1983, p. 259)

But mathematicians have always been drawn to the idea that mathematics is a fundamentally *descriptive science*. The set theorist Mirna Džamonja recently stated the view this way:

> I think that the observable reality is only a small part of the actual reality. This view is supported by developments in various physical sciences, of course, where one can talk about many objects that cannot or have not been observed so far. In mathematics, specifically in set theory, this means that for me objects such as ω_1 or $\aleph_\omega$ are in no way less real than the number 3, which is in turn no less real than a table or a chair. I am therefore a very strong platonist, to the point that I cannot even entertain having a different view. The axioms I view as an approximation of reality, and the fact that they do not (provably) describe the whole of the reality, is not surprising to me. The opposite would have been a serious surprise. (Personal communication.)

It is possible that Gödel's remarks and writings on Platonism attracted more attention of (analytic) philosophers than on anything else he ever wrote. His Platonism takes various forms and indeed Gödel formulated his own position differently at different times. He often expressed this as the view that mathematics is contentual:

> Logic and mathematics—like physics—are built up on axioms with a real content and cannot be explained away. The presence of this real content is seen by studying number theory. We encounter facts which are independent of arbitrary conventions. These facts must have a content because the consistency cannot be based on trivial facts... (Wang 1996, remark 7.1.4)

On other occasions Gödel meant Platonism to refer to the idea that precisely stated mathematical conjectures, such as the continuum hypothesis,[4] are either true or false. He sometimes used the term Platonism (or, interchangeably, realism) to indicate the idea that mathematical truth is itself objective; and on still other occasions he expressed himself in an ontological vein:

> [By the Platonistic view I mean the view that] mathematics describes a non-sensual reality, which exists independently both of the acts and the dispositions of the human mind and is only perceived, and probably perceived very incompletely, by the human mind. (Gibbs Lecture, Gödel 1995, pp. 304–323)

Somewhere toward the mid-1950s Gödel's Platonism became qualified and complicated by a regulative principle called *epistemological parity*, a fact that has gone largely unnoticed in the literature on Gödel's Platonism (but see van Atten and Kennedy 2003). This is the idea that, regarding physical objects on the one hand and abstract or mathematical objects on the other, from the point of view of what we know about them, there is no reason to be more (or less) committed to the existence of one than of the other:

> It seems arbitrary to me to consider the proposition "This is red" an immediate datum, but not so to consider the proposition stating modus ponens. (Gödel 1990, p. 347)[5]

The interpretive predicament posed by epistemological parity aside, it is well known that philosophers mounted a sustained attack on Platonism throughout the twentieth century, if not earlier,[6] the discussion becoming acute during the so-called Grundlagenstreit of the early twentieth century and the ensuing rise of the various foundational schools. Of course those discussions were not about Platonism explicitly; but Platonism, as Bernays and subsequent generations of philosophers called it, was, for the most part, the underlying issue at stake in many of those discussions.

---

[4] The continuum hypothesis states that the cardinality of any infinite set of real numbers is either that of the natural numbers, or that of the full set of real numbers. It is independent of the axioms of set theory: its consistency with those axioms was shown by Gödel in 1934; the consistency of its negation was shown by P. J. Cohen in 1963.

[5] The principle, in another form, is stated already in Gödel's 1944 essay on Russell: "It seems to me that the assumption of such objects is quite as legitimate as the assumption of physical bodies and there is quite as much reason to believe in their existence" (Gödel 1990, p. 128). The principle preoccupied Gödel through the 1960s, for example, the idea occurs in some form in a note to himself in the folder titled Phil[osophische] Varia (mostly after 1961): "It should also be noted that even a statement like 'this is red' *if there is to be a valid motive for making it* presupposes that something besides the independent sense experience is given." [GN folder 12/43, 060572; emphasis ours.] A view very similar to epistemological parity surfaces also in Tait's "On the Platonism of mathematics," in Tait (2005) and in some form in Burgess's essay "Why I am not a nominalist," in Burgess (2008).

[6] For example in some form in debates between Kronecker and Cantor.

As of this writing, nominalistic reconstructions of mathematical theories remain very attractive to philosophers. And if the motivation behind such reconstructions runs counter to the mathematician's understanding of her own practice, or if those reconstructions suffer from *awkwardness*, that is if they reconfigure the enterprise to such an extent as to be unrecognizable to its practitioners, then so be it. The mathematician is left with an uninterpreted world view, and the nominalist is charged with irrelevance by the very community whose scientific practice he set out to explain.

Now nominalism, as John Burgess notes,[7] is a big subject, but it is not ours. We wish to say something else here; in particular, we wish to ask the question, what does Gödel's philosophical legacy amount to?

3 Gödel's legacy: decidability in set theory

To assert, as Gödel did, that problems like the continuum hypothesis are solvable, is to embroil oneself in one of the main, if not the main, philosophical dispute of the field.[8] For now we only observe that a nontrivial number of set theorists are by and large somewhat reluctant to give up bivalence, and indeed there is at present a vibrant philosophical literature in this area of foundations of set theory, tracking what can only be described as a spectacular collection of results which have been obtained by set theorists over the last few decades.[9]

Set forcing is the main cause of variability in set theory,[10] but it turns out that the understanding of the set-theoretic universe has advanced to such a point that we now know that some of this variability can be disabled. For example, an axiom like Projective Uniformization is true "across the multiverse" (see Chapters 8 and 9). That is, it is true in "all" set-theoretic universes (i.e., those universes which are either (set) forcing extensions of *V*, or those of which *V* is a forcing extension), assuming that one, and hence all of these universes, have enough large cardinals.[11] In fact, by a result of

[7] "Being explained away," in Burgess (2008).
[8] For example, Tait has recently said about this project (see Tait 2008):

> For me, the most important open problem in philosophy of mathematics is in foundations of mathematics, and that is the search for new axioms of set theory, which means, too, the search for grounds for accepting them.

[9] See for example the contributions of Feferman, Friedman, Maddy and Steel (2000). See also Maddy (1998a) as well as the more recent Maddy (2009), Woodin (2001a, 2001b, 2002), Koellner (2009) and Bagaria (2006).
[10] Or at least one of them, depending on one's point of view.
[11] This is a series of results due to Hugh Woodin, D. A. Martin and John Steel.

Hugh Woodin and assuming large cardinals again, this also holds true for *all*
Σ^2_1 statements, that is, these are generically absolute,[12] assuming the
continuum hypothesis. Finally, and again under the assumption of enough
large cardinals, the theory of $L[\mathbb{R}]$ is generically absolute.[13]

How far this form of decidability will extend is not known. As of this
writing, the continuum problem remains undecided in this sense, that is, it
is not necessarily true across the multiverse, and cannot be if the multiverse
is based on set forcing.

In addition to the generic absoluteness results, forcing axioms in the
form of maximality principles, which Gödel advocated, actually decide the
value of the continuum in the direction of what was for a time Gödel's
suggested value, $\aleph_2$. The so-called *core model program* searches for canonical *L*-like inner models, that is, models built up from small "manageable"
pieces, which decide the continuum hypothesis and many other canonical
statements. Other results fix the theory of certain canonical structures in
the presence of large cardinals. Indeed, the large cardinals not only decide
individual statements, they introduce a conceptual coherence into the
whole set-theoretic universe, just as Gödel predicted they would.

Gödel's careful pursuit of his program in the foundations of mathematics and set theory was driven by decidability, as we have said. What is
striking about contemporary set-theoretic practice, as it happens – whether
this is due to Gödel's advocacy or whether this happened on its own – is
that much work in set theory nowadays also revolves around, if it is not
explicitly driven by, decidability. Gödel's unbending commitment to
decidability did involve a wider, elevated view of human rationality – as
did Hilbert's, for example, whose concept even of human dignity itself,
was linked to decidability in mathematics (Hilbert 1930a).

What one can say is this: the simple commitment to "keeping mathematics as it is," as we have called it; to preserving mathematical practice *in
its original form* by extending to certain natural set-theoretic questions the
kind of decidability one is able to take for granted in the rest of mathematics, was, and is still, the goal. And whether or not it will be conclu-
sively established that set-theoretic variability is here to stay; whether or
not it will become clear that the mathematician really does live in a
"multiverse" rather than a universe, Gödel's program for large cardinals,
and much of the philosophical analysis that was activated by and around it,
remains the essential inheritance of every working set theorist.

[12] Meaning that their truth value cannot be changed by set forcing.
[13] This is also due to Woodin. For proofs of this and of Σ^2_1-absoluteness, see Larson (2004).

PART I

Gödel on intuition

CHAPTER 2

Intuitions of three kinds in Gödel's views on the continuum

John P. Burgess

Gödel's views on mathematical intuition, especially as they are expressed in his well-known article on the continuum problem,[1] have been much discussed, and yet some questions have perhaps not received all the attention they deserve. I will address two here.

First, an exegetical question. Late in the paper Gödel mentions several consequences of the continuum hypothesis (CH), most of them asserting the existence of a subset of the straight line with the power of the continuum having some property implying the "extreme rareness" of the set.[2] He judges all these consequences of CH to be implausible. The question I wish to consider is this: *What is the epistemological status of Gödel's judgments of implausibility supposed to be?* In considering this question, several senses of "intuition" will need to be distinguished and examined.

Second, a substantive question. Gödel makes much of the experience of the axioms of set theory "forcing themselves upon one as true," and at least in the continuum problem paper makes this experience the main reason

---

[1] First version, Gödel (1947); second version in Benacerraf and Putnam (1st edition, 1964, pp. 258–273), reprinted in Benacerraf and Putnam (1983), pp. 470–485. *Collected Works*, edited by Feferman *et al.*, reprints all works published by Gödel during his lifetime in volumes I (1986) and II (1990), a substantial selection from his *Nachlaß* in volume III (1993), and from his correspondence in volumes IV and V (2003a, 2003b). Both versions of the continuum problem paper are reprinted in volume II, pp. 176–187 and 254–270. Quotations here will be from the second version, and the pagination in citations will be that of the *Collected Works*.

[2] Gödel in two paragraphs in the first version (Gödel 1947, pp. 523–524) mentions seven consequences of CH in all, but the last one mentioned in the first paragraph follows almost immediately from the first one mentioned in the second paragraph, and hardly needs to be counted separately. The last two are rather special, and I will set them aside. The remaining four are all "extreme rarity" results. For the *cognoscenti*, two pertain to Baire category and being topologically small, while two pertain to Lebesgue measure and being metrically small. The "extreme rarity" properties involved are these:

   a universalized version of the property of being first category
   a universalized version of the property of having measure zero
   having countable intersection with any first category set
   having countable intersection with any measure zero set

The second version of the paper (Gödel 1990, p. 263) drops the last of these.

for positing such a faculty as "mathematical intuition." After several senses of "intuition" have been distinguished and examined, however, I wish to address the question: *In order to explain the Gödelian experience, do we really need to posit "mathematical intuition," or will some more familiar and less problematic type of intuition suffice for the explanation?* I will tentatively suggest that Gödel does have available grounds for excluding one more familiar kind of intuition as insufficient, but perhaps not for excluding another.

1   Geometric intuition

In the broadest usage of "intuition" in contemporary philosophy, the term may be applied to any source (or in a transferred sense, to any item) of purported knowledge not obtained by conscious inference from anything more immediate. Sense-perception fits this characterization, but so does much else, so we must distinguish *sensory* from *nonsensory* intuition. Narrower usages may exclude one or the other. Ordinary English tends to exclude sense-perception, whereas Kant scholarship, which traditionally uses "intuition" to render Kant's "Anschauung," makes sense-perception the paradigm case.[3]

If we begin with sensory intuition, we must immediately take note of Kant's distinction between *pure* and *empirical* intuition. On Kant's idealist view, though all objects of outer sense have spatial features and all objects of outer and inner sense alike have temporal features, space and time are features only of things as they appear to us, not of things as they are in themselves. They are forms of sensibility which we impose on the matter of sensation, and it is because they come from us, rather than from the things, that we can have knowledge of them in advance of interacting with the things. Only empirical, *a posteriori* intuition can provide specific knowledge of specific things in space and time, but pure intuition, spatial and temporal, can provide *a priori* general knowledge of the structure of space and time, which is what knowledge of basic laws of three-dimensional Euclidean geometry and of arithmetic amounts to.

Or so goes Kant's story, simplified to the point of caricature. Kant claimed that his story alone was able to explain how we are able to have the a priori knowledge of three-dimensional Euclidean geometry and of arithmetic that

---

[3] Translators of ordinary, nonphilosophical German, would perhaps most often render "Anschauung" as "view." The use of "intuition" for sense-perception conforms less well to the ordinary English meaning of the word than to its Latin etymology (from *intueri*, "to look").

we have. But as is well known, not long after Kant's death doubts arose whether we really do have any such a priori knowledge in the case of three-dimensional Euclidean geometry, and later doubts also arose as to whether Kant's story is really needed to explain how we are able to have the a priori knowledge of arithmetic that we do have. Gödel has a distinctive attitude towards such doubts.

As a result of developments in mathematics and physics from Gauß to Einstein, today one sharply distinguishes mathematical geometry and physical geometry; and while the one may provide a priori knowledge and the other knowledge of the world around us, neither provides a priori knowledge of the world around us. Mathematical geometry provides knowledge only of mathematical spaces, which are usually taken to be just certain set-theoretic structures. Physical geometry provides only empirical knowledge, and is inextricably intertwined with empirical theories of physical forces such as electromagnetism and gravitation.

And for both mathematical and physical geometry, three-dimensional Euclidean space no longer has any special status. For mathematical geometry, Euclidean space is simply one of many mathematical spaces. For physical geometry, Euclidean space is no longer thought to be a good model of the world in which we live and move and have our being. Already with special relativity, physical space and time are merged into a four-dimensional physical spacetime, so that it is only relative to a frame of reference that we may speak of three spatial dimensions plus a temporal dimension. With general relativity, insofar as we may speak of space, it is curved and non Euclidean, not flat and Euclidean; and a personal contribution of Gödel's to twentieth century physics was to show that, furthermore, insofar as we may speak of time, it may be circular rather than linear.[4]

The Kantian picture thus seems totally discredited. Nonetheless, while Gödel holds that Kant was wrong on many points, and above all in supposing that physics can supply knowledge only of the world as it appears to us and not as the world really is in itself, still he suggests that Kant may have been right about one thing, namely, in suggesting that time is a feature only of appearance and not of reality.[5]

[4] That is, there are solutions to the field equations of general relativity in which there are closed time-like paths (Gödel 1949a). (The existence of such paths, a feature of some but not all of Gödel's models, is generally considered less significant for physics than the "rotating universe" feature of all his models.)

[5] Gödel (1949b) argues that while the conclusion that time is subjective and not objective is suggested already by special relativity, the case is not conclusive without his own results in general relativity.

As for intuition, again there is a mix of right and wrong. Gödel writes:

> Geometrical intuition, strictly speaking, is not mathematical, but rather a priori physical, intuition. In its purely mathematical aspect our Euclidean space intuition is perfectly correct, namely it represents correctly a certain structure existing in the realm of mathematical objects. Even physically it is correct "in the small." (Gödel 2003a, pp. 453–454)[6]

Elaborating, let us reserve for the pure intuition of space (respectively, of time) "in its physical aspect" the label *spatial* (respectively, *temporal*), intuition, and for the same pure intuition "in its mathematical aspect" let us reserve the label *geometric* (respectively, *chronometric*) intuition. Gödel's view, recast in this terminology, is that spatial intuition is about the physical world, but is only locally and approximately correct, while geometric intuition is globally and exactly correct, but is only about a certain mathematical structure. It would be tempting, but it would also be extrapolating beyond anything Gödel actually says, to attribute to him the parallel view about temporal *versus* chronometric intuition.

If geometric intuition "in its mathematical aspect" is "perfectly correct," can it help us with the continuum problem? The question arises because the continuum hypothesis admits a geometric formulation, thus:

> Given two lines X and Y in Euclidean space, meeting at right angles, say that a region F in the plane they span *correlates* a subregion A of X with a subregion B of Y if for each point x in A there is a unique point y in B such that the point of intersection of the line through x parallel to Y and the line through y parallel to X belongs to F, and similarly with the roles of A and B reversed. Say that a subregion B of Y is *discrete* if for every point y of B, there is an interval of Y around y containing no other points of B. Then for any subregion A of X, there is a region correlating A either with the whole of the line Y or else with a discrete subregion of Y.

Furthermore, it is not just the continuum hypothesis but many other questions that can be formulated in this style.[7] Among such questions are the problems of descriptive set theory whose status Gödel considers briefly at the end of his monograph on the consistency of the continuum hypothesis.[8]

---

[6] Letter to Marvin Jay Greenberg, October 2, 1973, in Gödel (2003a), pp. 453–454. It is followed by a reply from Greenberg and a draft of a reply to that by Gödel, expressing doubt about the notion of any *non* Euclidean geometric intuition. It is more than possible that Greenberg does not understand "intuition" in the same sense as Gödel, but in the sense discussed in Section 3 of the present chapter.

[7] As explained, with figures, in §II.A.5.b of Burgess and Rosen (1997), in connection with "geometric nominalism."

[8] Gödel (1940), p. 67, note 1. Actually, this note mentions explicitly just the existence of a projective well-ordering of the real numbers in order type ω_1, only alluding to and not discussing further implications. For a filling in of Gödel's sketch and explicit treatment of further implications see Addison (1959).

Can geometric intuition help with any of these problems? More specifically, can Gödel's implausibility judgments about the "extreme rareness" results that follow from CH be regarded as geometric intuitions? Some more background will be needed before this question can be answered.

Gödel's student years coincided with the period of struggle – Einstein called it a "frog and mouse battle" – between Brouwer's intuitionism and Hilbert's formalism. It is rather surprising, given the developments in mathematics and physics that tended to discredit Kantianism, that the two rival schools both remained Kantian in outlook. Thus Brouwer describes his intuitionism as "abandoning Kant's apriority of space but adhering the more resolutely to the apriority of time,"[9] while Hilbert proposes to found mathematics on spatial intuition, treating it as concerned with the visible or visualizable properties of visible or visualizable symbols, strings of strokes.[10]

Hans Hahn, Gödel's nominal dissertation supervisor and a member of the Vienna Circle, wrote a popular piece alleging the bankruptcy of intuition in mathematics,[11] and thus by implication separating himself, like a good logical positivist, from both the intuitionist frogs and the formalist mice. Hahn alludes to the developments in mathematics and physics culminating in relativity theory as indications of the untrustworthiness of intuition, but places more weight on "counterintuitive" discoveries such as Weierstraß's curve without tangents and Peano's curve filling space.[12] Do such counterexamples show that geometric intuition is not after all "perfectly correct"?

---

[9] Brouwer (1913); reprinted in Benacerraf and Putnam (1983), pp. 66–77, with the quoted passage on p. 69. The paper dates from between the discovery of special relativity and that of general relativity. Kant's views on time as regards *outer* sense seem discredited by special relativity and the discovery that the temporal order of distant events is in general not absolute but relative to a frame of reference. But it is clear from the continuation of the passage that Brouwer is speaking of adhering to Kant's views on time only as regards *inner* sense. If those views, too, are threatened by developments in physics, it is by Gödel's results in general relativity.

[10] More precisely, Hilbert proposes to found *finitist* mathematics in this way; but finitist mathematics is for him the only "real" or *inhaltlich* mathematics. Charles Parsons has objected that though Hilbert regarded exponentiation as a legitimate operation of finitist arithmetic on a par with addition, there is a crucial difference; see Parsons (2008), especially chapter 7, "Intuitive arithmetic and its limits." The objection of Parsons is that while addition as an operation of strings of strokes can be visualized as juxtaposition, exponentiation seems to be visualizable only as a *process* rather than an *object*. But it remains that Hilbert's professed orientation, despite his deep interest in general relativity, is still quasi- or neo-Kantian to the same degree as Brouwer's. Of course, Hilbert does not make mathematics depend on geometric intuition in the way that Frege was driven to do after the collapse of his logicist program in contradiction: he does not revert to Newton's conception of real numbers as abstracted ratios of geometric properties, whose basic laws are to be derived from theorems of Euclidean geometry.

[11] Hahn (1956). Originally a lecture in German, it is very well known in the English speaking world from its appearance in print in English, no translator is named, in James R. Newman's anthology.

[12] The "counterintuitiveness" of these examples has been disputed by Benoît Mandelbrot (1977). His appears, however, to be a minority view.

Gödel in effect insists that there is no real "crisis in intuition" while conceding that there is an apparent one. Thus he writes:

> One may say that many of the results of point-set theory ... are highly unexpected and implausible. But, true as that may be, still ... in those instances (such as, *e.g.*, Peano's curves) the appearance to the contrary can in general be explained by a lack of agreement between our intuitive geometrical concepts and the set-theoretical ones occurring in the theorems. (Gödel 1990, p. 263)[13]

The appearance of paradox results from a gap between the technical, set-theoretic understanding of certain terms with which Weierstraß, Peano, and other discoverers of pathological counterexamples were working, and the intuitive, geometric understanding of the same terms.

Presumably the key term in the examples under discussion is "curve." The technical, set-theoretic concept of curve is that of a continuous image of the unit interval. The intuitive, geometric concept of curve is of something more than this, though unfortunately Gödel does not offer any explicit characterization for comparison. Unfortunately also, Gödel does not address directly other "counterintuitive" results in the theory of point-sets, where presumably it is some term other than "curve" that is associated with different concepts in technical set-theory and intuitive geometry.[14] Thus he leaves us with little explicit indication of what he takes the intuitive geometric concepts to be like.

But to return to his basic point about the divergence between intuitive geometric notions and technical set-theoretic notions, it is precisely on account of this divergence, and not because of any unreliability of geometric intuition in its proper domain, that Gödel is unwilling to appeal to geometric intuition in connection with the continuum problem. Gödel explicitly declines for just this reason to appeal to geometric intuition in opposition

---

[13] The importance of this passage has been noted by both of the commentators whose work has most influenced the present paper, Penelope Maddy and D. A. Martin, in their papers cited below. (Maddy in particular explicitly reaches the conclusion stated in the last two sentences of the present section.)

[14] For instance, despite his ringing endorsement of the axiom of choice as in all respects equal in status to the other axioms of set theory (Gödel 1990, p. 255 footnote 2), he does not discuss one of its most notorious geometrical consequences, the Banach–Tarski paradox, and this even though he cites the paper in which the word "paradox" was first applied to the Banach–Tarski result (Blumenthal 1940). The absence of an explicit Gödelian treatment of this example is especially regrettable because one suspects that what Gödel would have said about this case, where "intuitions" contrary to set-theoretic results seem to be based on the assumption that any region of space must have a well-defined volume, might well extend to the "intuitions" appealed to in Chris Freiling's (1986) infamous argument against the continuum hypothesis, which commentators have seen as assuming that any event must have a well-defined probability.

to one of the easier consequences of the continuum hypothesis derived in Sierpinski's monograph on the subject.[15] The consequence in question is that the plane is the union of countably many "generalized curves" or graphs of functions $y = f(x)$ or $x = g(y)$.[16] This may appear "highly unexpected and implausible," but this notion of "generalized curve" is even further removed from the intuitive, geometric notion of curve than is the notion of a curve as any continuous image of the unit interval.[17] Thus no help with the continuum problem is to be expected from geometric intuition. We must conclude that Gödel's implausibility judgments are not intended as reports of geometric intuitions. They must be something else.

2 Rational intuition

It is time to turn to nonsensory as opposed to sensory intuition, which will turn out to be a rather heterogeneous category. Let us proceed straight to the best-known passage in the continuum problem paper, which speaks of "something like a perception" even of objects of great "remoteness from sense experience":

> But, despite their remoteness from sense experience, we do have something like a perception also of the objects of set theory, as is seen from the fact that the axioms force themselves upon us as true. I don't see any reason why we should have less confidence in this kind of perception, i.e., in mathematical intuition, than in sense perception... (Gödel 1990, p. 267)[18]

The passage is as puzzling as it is provocative.

Almost the first point Charles Parsons (2008) makes in his recent extended discussion of the usage of the term "intuition" in philosophy of mathematics is that it is crucial to distinguish intuition *of* from intuition *that*. One may, for instance, have an intuition *of* a triangle in the Euclidean

[15] See Gödel (1990), p. 269, the second of four numbered remarks at the beginning of the supplement added to the second version, for Gödel's remarks on Waclaw Sierpinski, *L'Hypothèse du Continu*. The particular consequence alluded to is among the *equivalents* of CH listed in the book, where it is named P_2. Gödel cites both the first edition (1934) and the second (1956).

[16] The continuum hypothesis implies that there is an ordering of the real numbers in which for each x there are only countably many y less than x. The axiom of choice allows us to pick for each x a function h_x from the natural numbers onto the set of such y. Then we may define functions $f_n(x) = h_x(n)$, and the graphs of these functions, plus their reflections in the diagonal $y = x$, plus the diagonal itself, give countably many "generalized curves" filling the plane.

[17] Even Mandelbrot's more expansive conception of what is intuitive seems to take in only F_σ or G_δ or anyhow low-level Borel sets (to which classifications his "fractals" all belong), not arbitrary "generalized curves."

[18] This passage comes from the supplement added to the second version of the paper.

plane without having an intuition *that* the sum of its interior angles is equal to two right angles.[19] Gödel, by contrast, seems in the quoted passage to leap at once and without explanation from an intuition *of* set-theoretic objects to an intuition *that* set-theoretic axioms are true. What is the connection supposed to be here? It is natural to think that perceiving or grasping set-theoretic *concepts* (set and elementhood) would involve (or even perhaps just consist in) perceiving or grasping *that* certain set-theoretic axioms are supposed to hold; but why should one think the same about perceiving set-theoretic *objects* (sets and classes)? After all, we have not just "something like" a perception but an outright perception of the objects of astronomy, but when we look up at the starry heavens above, no astronomical axioms force themselves upon us as true.[20]

The passage is the more puzzling because one can find, even within the same paper, passages where Gödel seems to distance himself from any claim to have intuition of mathematical objects individually:

> For someone who considers mathematical objects to exist independently of our constructions and of our having an intuition of them individually, and who requires only that the general mathematical concepts must be sufficiently clear for us to be able to recognize their soundness and the truth of the axioms concerning them, there exists, I believe, a satisfactory foundation for Cantor's set theory in its whole original extent and meaning, namely, axiomatics of set theory interpreted in the way sketched below. (Gödel 1990, p. 258)[21]

There is scarcely room for doubt that Gödel is thinking of himself as such a "someone." If so, then he seems to be insisting only on an understanding of "general mathematical concepts," not a perception of individual mathematical objects.

---

[19] See Parsons (2008), p. 8. This book has had a greater influence on the present paper than will be evident from my sporadic citations of it. Inversely, Parsons holds, as a consequence of his structuralism, that we can have an intuition *that* every natural number has a successor, though we have no intuition *of* natural numbers. See Parsons (2008), §37, "Intuition of numbers denied," pp. 222–224.

[20] The most plausible account to date of how and in what sense we might be said to perceive sets is that of Penelope Maddy (1990), especially chapter 2, "Perception and intuition." But on this account set-theoretic perception is mainly of small sets of medium-sized physical objects, just as sense-perception is mainly of medium-sized physical objects themselves. The theoretical extrapolation to infinite sets then seems to have the same status as the theoretical extrapolation to subvisible physical particles, and this would seem to leave the axiom of infinity with the same status as the atomic hypothesis: historically a daring conjecture, which by now has led to so much successful theorizing that we can hardly imagine doing without it, but still not something that "forces itself upon us."

[21] The passage comes from §3 of the paper and leads into Gödel's exposition of the cumulative hierarchy or iterative conception of set (which is what the phrase "the way sketched below" in the quotation refers to).

The general view of commentators, expressed already many years ago by William Tait (1986),[22] is that one should not, on the strength of the puzzling passage, attribute to Gödel the view that we have a kind of ESP by which we can observe the elements of the set-theoretic universe. Tait points to clues to what Gödel may mean by "something like a perception of the objects of set theory" in Gödel's statements in adjoining passages to the effect that (i) even in the case of sense-perception we do not immediately perceive physical objects but form ideas of them on the basis of what we do immediately perceive, and (ii) the problem of the existence of mathematical objects is an exact replica of the problem of the existence of physical objects. Tait does not explicitly say what conclusions about what Gödel meant should be drawn from these passages, except to repudiate the ESP interpretation.

The conclusion one might think suggested would be this. The experience of the axioms forcing themselves upon us is like the experience of receiving sense-impressions, and inferring the set-theoretic objects from the experience of the axioms forcing themselves upon is like inferring physical objects from sense-impressions. But there is a well-known problem with such a view. From sensations we infer material bodies as their *causes*, but if we are to avoid claims of ESP, we must not suppose that the sets can be inferred as *causes* of our feeling the axioms forced upon us. They are presumably inferrable, once the axioms have forced themselves upon us, only as things behaving as the axioms say sets behave; and the problem is that this will not distinguish the genuine sets from the elements of any isomorphic model, a point familiar from discussions of structuralism in philosophy of mathematics.

D. A. Martin (2005) has looked closely at the puzzling passage about "something like a perception of the objects of set theory" with a structuralist point of view in mind, denying like other commentators that Gödel is committed to the perceptibility of individual sets, and if I read him aright suggesting that Gödel may be speaking of the perception of the *structure* of the set-theoretic universe, rather than its elements.[23] The interpretation of Gödel as a structuralist may, however, seem anachronistic to some. A slightly different interpretation is available. For in the course of his study Martin collects textual evidence from a variety of Gödelian sources to show that Gödel does not, as Frege does, think of "objects" and "concepts" as nonoverlapping

[22] Tait (1986), see note 3, pp. 364–365.

[23] I will not be doing justice to this study, which would require extended discussion of structuralism. In particular I will not be discussing what real difference, if any, there would be between perceiving the *structure* of the universe of sets as Martin understands it and perceiving the *concept* of set as Gödel understands concepts. (Both are clearly different from perceiving the individual sets that occupy positions in the structure and exemplify the concepts.)

categories, but rather thinks of concepts as a species within the genus of objects. This makes it at least conceivable that when Gödel speaks of a perception of the objects of set theory, he has in mind perception of the concepts of set theory, and that it does not seem as odd to him as it would to some of us to call these concepts "objects."

Parsons, too, seems to take Gödel to be including concepts among the "objects of set theory" in the passage under discussion.[24] In what follows I will take it that for Gödel we have something like a perception of the concept of set, bringing with it (or even perhaps just consisting in) axioms forcing themselves upon us. Such a reading makes Gödel an adherent of the view that there is a faculty resembling sense-perception but directed towards abstract ideas rather than concrete bodies. Commentators call such a faculty *rational* intuition.[25]

Rational intuition as applied specifically to mathematical concepts may be called *mathematical* intuition. Mathematical intuition as applied specifically to set-theoretic concepts may be called *set-theoretic* intuition. The geometric and chronometric intuitions encountered in the preceding section really should be reclassified as forms of mathematical intuition. Gödel does not tell us much about forms of mathematical intuition other than set-theoretic and geometric, let alone about forms of rational intuition other than mathematical; nor does he consider forms of nonsensory intuition other than rational (of which more below).

Belief in such a faculty as rational intuition is hardly original with or unique to Gödel. Thus Diogenes Laertius relates the following tale of an exchange between his namesake Diogenes the Cynic and Plato:

> As Plato was conversing about Ideas and using the nouns "tablehood" and "cuphood," he [the Cynic] said, "Table and cup I see; but your tablehood and cuphood, Plato, I can nowise see." "That's readily accounted for," said Plato, "for you have eyes to see the visible table and cup; but not the understanding by which ideal tablehood and cuphood are discerned." (Diogenes Laertius, Book VI, p. 55)[26]

---

[24] See Parsons (1995a), where he discusses the passage at issue on p. 65. In helpful comments on a preliminary version of the present study, Parsons remarks, "One piece of evidence ... is that Gödel frequently talks [elsewhere] of perception of concepts but hardly at all about perception or intuition of sets. It may be that any perception of sets that he would admit is derivative from perception of concepts," here alluding to the suggestion made in footnote 43 of the cited paper that those sets, such as the ordinal ω, that are individually definable may be "perceived" by perceiving the concepts that identify them uniquely – though, of course, what it identifies uniquely is really only the position of the ordinal in the set-theoretic universe.

[25] See Parsons (2008), §52, "Reason and 'rational intuition'" for some healthy skepticism about the appropriateness of this traditional term.

[26] Quotation is from Diogenes Laertius (1925).

Nowadays the label "Platonist" is bandied about rather loosely in philosophy of mathematics, but for Gödel the label really fits.

Ever a Platonist in this sense, Gödel became first something of a Leibnizian and then more of a Husserlian as he sought a home in a systematic philosophy for his basic belief in rational intuition. Gödel reportedly took up Husserl between the appearance of the first version and the second version of the continuum problem paper. Commentators more familiar with Husserl and phenomenology than I am have seen evidence of Husserlian influence in some of the new material added to the second version.[27] The suggestion seems to be that the study of phenomenology may have led Gödel to put less emphasis on the supposed independent existence of mathematical objects, and more on other respects in which what I am calling rational intuition of concepts is supposed to resemble sensory intuition of objects.

Some respects come easily to mind, and can be found mentioned more or less explicitly in Gödel. Like sense-perceptions, rational intuitions are not the product of conscious inference, being observations rather than conclusions. Like sense-perceptions, rational intuitions constrain what we can think about the items they are perceptions or intuitions *of*, since we must think of those items as having the properties we observe them to have. Like sense-perceptions, rational intuitions seem open-ended, seem to promise a series of possible further observations. Like sense-perception, rational intuition can be cultivated, since through experience one can develop abilities for closer and more accurate observation.

One important point of resemblance needs to be added to the list: like sense-perceptions, rational intuitions are fallible, and errors of observation sometimes lead us astray. Gödel emphasizes this feature more in his paper on Russell, where he naturally has to say something about the paradoxes, than in the one on Cantor. He describes Russell as

---

[27] In particular, Kai Hauser in a talk at the 2009 NYU conference in philosophy of mathematics cited as evidence of Husserlian influence the following somewhat concessive passage (which has also drawn the attention of earlier commentators):

> However, the question of the objective existence of the objects of mathematical intuition ... is not decisive for the problem under consideration here. The mere psychological fact of the existence of an intuition which is sufficiently clear to produce the axioms of set theory and an open series of extensions of them suffices to give meaning to the question of the truth or falsity of propositions like Cantor's continuum hypothesis. (Penultimate paragraph of the supplement, Gödel 1990, p. 268)

... bringing to light the amazing fact that our logical intuitions (*i.e.*, intuitions concerning such notions as: truth, concept, being, class, etc.) are self-contradictory. (Gödel 1944, reprinted in Benacerraf and Putnam 1983)[28]

Apparently, then, the mere fact that Russell found a contradiction in Frege's system would not in itself necessarily count for Gödel as conclusive evidence that Frege did not have a genuine rational intuition in favor of his Law V. A similar remark would presumably apply to the well-known minor fiasco in Gödel's declining years, when he proposed an axiom intended to lead to the conclusion that the power of the continuum is $\aleph_2$ but actually implying that it is $\aleph_1$.[29]

It may be mentioned that if rational intuition is really to be analogous to sensory intuition, then there must not only be cases where rational intuition is incorrect, but also cases where it is indistinct, like vision in dim light through misty air. And there is something like dim, misty perception of a concept in Gödel. For instance, Gödel seems to see, looming as in a twilight fog beyond the rather small large cardinal axioms he is prepared to endorse (inaccessible and Mahlo cardinals), further principles or maybe one big principle that would imply the existence of much larger cardinals, but that he is not yet in a position to articulate.[30]

The crucial philosophical question about rational intuition, however, is not how bright or dim it is; nor even how reliable or treacherous it is; nor yet how long or short the list of analogies with sense-perception is; and least of all whether "rational intuition" is right or wrong as a label for it. The crucial philosophical question is simply whether there is any real need to posit a special intellectual faculty in order to account for the experiences of the kind Gödel describes, where axioms "force themselves upon us,"

---

[28] "Russell's mathematical logic," in Benacerraf and Putnam (1983), pp. 221–232, with the quoted passage on pp. 215–216. Gödel's "Platonism" or "realism" is nearly as evident in this work as in the continuum problem paper. Parsons, in correspondence, while agreeing that Gödel acknowledged the fallibility of rational intuition, and emphasizing that in so acknowledging Gödel was departing from the earlier rationalist tradition, nonetheless warns against reading too much into the quoted passage, on the grounds that Gödel's usage of "intuition" may have been looser than at the time of the Russell paper than it later became.

[29] The documents (two notes and an unsent letter by Gödel), and an informative discussion of the unedifying episode by Robert Solovay, can be found in Gödel (1995), pp. 405–425. Another example of the fallibility of intuition may perhaps be provided by the fact mentioned by Solovay, that the pioneering descriptive set theorist Nikolai Luzin, who disbelieved CH, connected his disbelief with "certainty" that every subset of the reals of size $\aleph_1$ is coanalytic. We now know, however, that assuming a measurable cardinal, if CH fails then *no* set of size $\aleph_1$ is coanalytic (since assuming a measurable cardinal, every coanalytic set is either countable or of the power of the continuum).

[30] His formulations, however, in Gödel (1990), p. 260, footnote 20 and the text to which it is attached, are rather cautious, and he mentions on the next page that "there may exist ... other (hitherto unknown) axioms."

or whether, on the contrary, such experiences can be explained in terms of faculties already familiar and less problematic. For there are other, more mundane, varieties of nonsensory intuition, and a skeptic might suspect that one or another of them is what is really behind Gödelian experiences.

There is, for instance, *linguistic* intuition. Linguistic intuitions are simply the more or less immediate judgments of competent speakers to the effect that such-and-such a sentence is or is not syntactically or semantically in order. In both scientific linguistics and philosophical analysis such intuitions provide the data against which syntactic or semantic rules and theories are evaluated. Even theorists who suppose that competent speakers arrive at their linguistic intuitions by unconsciously applying syntactic or semantic rules do not suppose that there is any psychoanalytic procedure to bring these unconscious rules to consciousness. The only way to divine what the rules must be is to formulate hypotheses, test them against the data that *are* conscious, namely, linguistic intuitions, then revise, retest, and so on until the dialectic reaches stable equilibrium.

Is familiar linguistic intuition enough to explain Gödelian experiences when axioms "force themselves upon us," or do we need to posit a more problematic rational intuition? Perhaps we should ask first just what the difference between appeal to one and appeal to the other amounts to. The two appeals seem to go with two different pictures, both starting from something like Gödel's exposition of the cumulative hierarchy or iterative conception of sets.[31]

On the linguistic picture, from that exposition and the meanings of the words in it we deduce by logic set-theoretic axioms, and then from these by more logic we deduce mathematical theorems. Since as competent speakers we know the meanings of the words in the exposition, and since we are finite beings, the meanings must themselves be in some sense finite. The mathematical theorems we can deduce are thus deducible by logic from a fixed finite basis.

On the rationalist picture, the only function of the original exposition is to get us to turn our rational intuition in the direction of the concept of set.

[31] In §3 "Restatement of the problem. . ." or in other expositions of the same kind, several of which can be found in §IV "The concept of set" of Benacerraf and Putnam (1983). Note, however, that two of the contributors there, George Boolos ("The iterative conception of set," pp. 486–502) and Charles Parsons ("What is the iterative conception of set?" pp. 503–529) in effect deny the reality of Gödelian experiences, deny that the axioms do "force themselves upon us." They do so also in other works (Boolos 1998; Parsons 2008, §55, "Set theory," pp. 338–342). In this chapter I will not debate this point, but will simply grant for the sake of argument that Gödel is right and in fact there occurs such a phenomenon as the axioms "forcing themselves upon one." The issue I wish to discuss is, granting that in fact such experiences occur, whether we need to posit rational intuition to explain their occurrence.

Once we perceive it, we can go back to it again and again and perceive more and more about it. Hence, though at any stage we will have perceived only finitely much, still we have access to a potentially infinite set of set-theoretic axioms, from which to deduce by logic mathematical theorems. The mathematical theorems we can deduce are thus not restricted to those deducible by logic from a fixed finite basis.[32]

Now it is a consequence of Gödel's first incompleteness theorem that deduction by first-order logical rules from a fixed finite basis of first-order nonlogical axioms will leave some mathematical questions unanswered, whatever the fixed finite basis may be. One cannot speak of strict entailments in connection with the kind of broad-brush picture painting we have been engaged in, but one can say that, in view of Gödel's result, the linguistic picture *tends to suggest* that there must be absolutely undecidable mathematical questions, while the rationalist picture *tends to suggest* that there need not be.[33]

Or perhaps that overstates the matter. On the one hand, since semantic rules are not directly available to consciousness, and definitions doing full justice to the conventional linguistic meaning of a word are not always easy to find – witness decades of attempts by analytic philosophers to define "S knows that p" – when accepted axioms fail to imply an answer to some question, it is conceivable that they simply fail to incorporate everything that is part of the conventional linguistic meaning of some key term, and that appropriate use of linguistic intuition may lead to new axioms. On the other hand, even if it is assumed we have a rational intuition going beyond linguistic intuition, training this intellectual vision once again on the key concept may not be enough to give an answer to a question not decided by accepted axioms, since presumably there are limits to the acuity of metaphorical as much as to literal vision. It remains, however, that in any specific case of a question left undecided by current axioms, the one picture tends to inspire pessimism and the other optimism about the prospects for finding an answer.

[32] The kind of view I am attributing to Gödel resembles the kind of view Tyler Burge attributes to Frege, see Burge (2005). Frege sometimes says that everyone has a grasp of the concept of number and sometimes says that even very eminent mathematicians before him lacked a sharp grasp of the concept of number. Burge proposes to explain Frege's speaking now one way, now the other, by suggesting that Frege distinguishes the kind of minimal grasp of the associated concept possessed by anyone who knows the fixed, conventional linguistic meaning of an expression, with the ever sharper and sharper grasp which not every competent speaker of the language, by any means, can hope to achieve.

[33] Something like the contrast I have been trying to describe was, I suspect, ultimately the issue between Gödel and Carnap, but examination of that relationship in any detail is out of the question here. A complication is that Gödel sometimes uses "meaning" related terms in idiosyncratic senses, so that he ends up saying that mathematics is "analytic" and thus *sounding* like Carnap, though he does not at all *mean* by "analytic" what Carnap would. Martin and Parsons both discuss examples of this usage.

That may be a reason to *hope* that the rationalist rather than the linguistic picture is the correct one, but have we any reason to *believe* it is? Gödel does not really address this question, but it seems clear to what evidence he would point, and what kind of claim he would have to make about it, namely, the claim that the standard axioms of set theory plus some large cardinals "force themselves upon us," *even though they are not strictly rigorously logically implied by the literal conventional linguistic meaning of his exposition* of the cumulative hierarchy or iterative conception of sets.

Readers may exercise their own intuitions to evaluate such a claim. For what it is worth, I myself do feel that, say, inaccessible cardinals are implied by the *spirit* but not the *letter* of Gödel's exposition. To me, deciding for or against inaccessibles seems a bit like a judge deciding one way or the other in a kind of case that was never anticipated by the legislature and which the literal meaning of the words of the applicable law does not settle unambiguously one way or the other. A decision in one direction may be in the spirit of the law and in the other contrary to it, even though it cannot be said that the letter of the law strictly implies the one or contradicts the other.

If all this is so, then the alleged instances of rational intuition that Gödel cites cannot be explained as instances of linguistic intuition. But explaining apparent rational intuitions as really linguistic intuitions is not the only alternative to recognizing a special faculty of rational intuition. For there may remain yet other kinds of intuition to be considered. After all, *something* has led Gödel to his implausibility judgments about "extreme rareness" results. It is certainly not linguistic intuition, and unless it can be claimed to be rational intuition, it must be something *else* that we have not yet considered.

Nothing Gödel says suggests that he takes his implausibility judgments to be clear rational intuitions. If they were, then he would presumably advocate the denials of the consequences of CH judged implausible as new axioms, comparable to the new axioms of inaccessible and Mahlo cardinals; and this he does not do. Nothing Gödel says even suggests that he has a dim, misty perception of any *potential* for new axioms out in the *direction* of these implausibility judgments.[34] The only directions from

---

[34] It would be very difficult to formulate any such new axiom about extreme rarity, since nothing is more common in point-set theory than to find that sets small in one sense are large in another. Right at the beginning of the subject comes the discovery of the Cantor set, which is small topologically (first category) and metrically (measure zero), but large in cardinality (having the power of the continuum). Another classic result is that the unit interval can be written as the union of a first category set and a measure zero set. See Oxtoby, (1971) for more information (The particular result just cited appears as Corollary 1.7, p. 5.) The difficulty of finding a rigorous formulation, however, is only to be expected with dim and misty rational intuitions.

which Gödel even hints that a solution to the continuum problem might be sought is from large cardinals or something of the sort,[35] and that remains the most important direction being pursued today.[36]

It may also be pointed out that, while we have seen Gödel speak of "mathematical intuition" in the passage quoted at the beginning of this section, he never applies the term "intuition" to his implausibility judgments.[37] We must conclude that Gödel's implausibility judgments are not intended as reports of rational intuitions. They must be something else.

3   Heuristic intuition

Gödel is interested in rational intuition as the source of axioms from which mathematical deductions can proceed, but he shows very little interest in the

[35] Here "something of the sort" may be taken to cover the suggestion of looking for some sort of maximal principle, made in footnote 23 (Gödel 1990, p. 262). Gödel also mentions (p. 261) the possibility of justifying a new axiom not by rational intuitions in its favor, but by verification of striking consequences. Gödel cites no candidate example and even today it is not easy to think of one, if one insists that the striking consequences be not just aesthetically pleasing, like the pattern of structural and regularity properties for projective sets that follow from the assumption of projective determinacy, but *verified*. The one case I can think of is Martin's proof of Borel determinacy (as a corollary of analytic determinacy) assuming a measurable cardinal before he found a more difficult proof without that assumption. And in this example the candidate new axiom supported is still a large cardinal axiom.

[36] To be sure, in the wake of Cohen's work, Levy and Solovay (1967) showed that no solution to the continuum problem is to be expected from large cardinal axioms of a straightforward kind. But the present-day Woodin program can nonetheless be considered as in a sense still pursuing the direction to which Gödel pointed. According to Woodin's talk at the 2009 NYU conference in philosophy of mathematics, one of the possible outcomes of that program would be the adoption of a new axiom implying (1) that power of the continuum is $\aleph_2$ and (2) that Martin's Axiom (MA) holds. (1) is something Gödel came, at least for a time, to believe (in connection with the unedifying square axioms incident alluded to earlier). (2) is shown by Martin and Solovay (1970), in the paper in which MA was first introduced (see especially, §5.3, "Is A true?" pp. 176–177), to imply many of the same consequences as CH. In particular, MA implies several of the consequences about extreme rarity that Gödel judges implausible, plus a modified version of another that Gödel might well have judged nearly equally implausible.

[37] The implausibility judgments are at least indirectly classified as "intuitions" by commentators. Martin and Solovay (1970, p. 176) contrast Gödel's opinion with their own "intuitions," thus:

> If one agrees with Gödel that [the extreme rareness results] are implausible, then one must consider [MA] an unlikely proposition. The authors, however, have virtually no intuitions at all about [the extreme rareness results]. . .

Martin (1976) refers to Gödel's judgments as "intuitions" as he expresses dissent from them, thus:

> While Gödel's intuitions should never be taken lightly, it is very hard to see that the situation is different from that of Peano curves, and it is even hard for some of us to see why the examples Gödel cites are implausible at all.

The usage of the commentators here is in conformity with the kind of usage of "intuition" in mathematics to be discussed in the next section; but it seems Gödel's usage is more restricted than that.

source of the deductions themselves. How are they discovered? As a creative mathematician of the highest power, Gödel will have known from personal experience, and far better than any commentator, that a deduction is not discovered by first discovering the first step, second discovering the second step, and so on. But he does not display anything like Hadamard's or Polya's interest in the psychology of discovery in the mathematical field, insofar as this pertains to discovery of proofs from axioms rather than of the axioms themselves.

Hadamard emphasizes the role of the unconscious, while Polya emphasizes the role of induction and analogy in mathematical proof-discovery and problem-solving. Both tend to make the thought-processes of the working mathematician rather resemble those of the empirical scientist.[38] Gödel in his reflections on mathematical epistemology takes little note of such matters. Nor does he take much note of the fact that the body of theorems of mathematics comes accompanied by a body of conjectures for which no rigorous proof has yet been found.[39] When mathematicians speak of "intuition," however, it is perhaps most often in connection with just these matters that Gödel neglects.

Thus it is said to be by intuition that one comes to suspect what the answer to a question must be before one finds the proof, or that a proof is to be sought in *this* rather than *that* direction. This use of "intuition" for the faculty of arriving at hunches, and "intuitions" for the hunches arrived at, is by no means confined to mathematics, but is probably closer to the ordinary sense of the word than any of the special senses of "intuition" considered by philosophers. In contrast to other kinds let me call this *heuristic* intuition.

There is, perhaps, less appearance of immediacy with heuristic than with sensory or linguistic intuitions. To be sure, no kind of intuition is immediate in an absolute sense. A great deal of processing goes on at a subconscious or "subpersonal" level during the very short interval between the exposure of the retina to light and the resulting visual experience, or between our

---

[38] Hadamard (1945), Polya (1954). The resemblance between mathematical and scientific methodology is most conspicuous in Polya's second volume, where the patterns of plausible inference Polya detects in mathematical thought closely resemble the rules of Bayesian probabilistic inference often cited in work on the epistemology of science. It is, however, difficult to view them as literal instances, since the Bayesians often require that all logicomathematical truths be assigned probability one.

[39] There are also principles for which we do not even have a rigorous statement, let alone a rigorous proof. Such is the case with the Lefschetz principle, or Littlewood's three principles, for instance. Rigorous formulations of *parts* of such principles are possible, but always fall short of their full content. The "rules of thumb" in set theory identified by Maddy (1988b) may also be considered to be of this type.

exposure to a sentence and our judgment that it is or is not good English. But none of this processing, even if one wants to say that it is in some sense a process of "inference," is the kind of inference that could be brought to consciousness.

By contrast, when one has a hunch, in mathematics or elsewhere, when "something makes one think that" such-and-such is the case, often it turns out to be possible after some effort to articulate, at least partially, what the *something* is, and then the heuristic intuition becomes no longer an intuition but a heuristic *argument*. Typically the premises in a heuristic argument are a mix, with some that have been rigorously proved and others that are merely plausible conjectures (that is to say, that are themselves heuristic intuitions), while the inferential steps are a mix of logically valid and merely plausible transitions (such as reasoning by induction or analogy). It is reasonable to suspect that in all cases some kind of unconscious reasoning by induction or analogy or some other form of merely plausible inference from mathematic facts with which one is familiar underlies heuristic intuitions.

Maddy (1988b) has usefully surveyed just about all the common heuristic arguments for and against CH.[40] Her survey, needless to say, includes the arguments against CH from Gödel's paper. The role of the implausibility judgments in that argument looks just like the role of heuristic intuitions in other arguments, and I am prepared to classify the implausibility judgments as heuristic intuitions, thus answering the exegetical question with which I began: What is the epistemological status of Gödel's judgments of implausibility supposed to be?

Of course, Gödel himself does not use this terminology, but the very title of the section of the paper where these implausibility judgments are advanced – "In what sense and in which direction may a solution of the continuum problem be expected?" – seems to indicate that Gödel understands his implausibility judgments to be just the sort of plausibility or implausibility judgments that mathematicians typically come up with when discussing a famous conjecture pro and con. The "only" difference is that mathematicians generally anticipate that the famous conjecture they

---

[40] See Maddy (1998b), §II.3, "Informed opinion," pp. 494–500. To give an example not in Maddy's collection, one might argue heuristically against the continuum hypothesis as follows. CH implies not only that all uncountable subsets of the line have the same number of elements, but also that all partitions of the line into uncountably many pieces have the same number of pieces. But even looking at very simple partitions (those for which the associated equivalence relation, considered as a subset of the plane, is analytic) with uncountably many pieces, we find what seem two quite different kinds. For some it can be proved that the number of pieces is exactly $\aleph_1$ and that there is no perfect set of pairwise inequivalent elements, while for others it can proved that there *is* such a perfect set and (hence) that the number of pieces is the power of the continuum. (Compare Srivastava (1988), chapter 5.)

are discussing will eventually be proved or disproved, whereas (so far as proof from currently accepted axioms is concerned) Gödel knows that CH cannot be disproved, and suspects that CH cannot be proved.[41]

Now if we are told that there is supposed to be a difference in status between rational intuitions in favor of large cardinals and merely heuristic intuitions against extremely rare sets of the power of the continuum, we may wonder how one is supposed to be able to *tell*, when one has a pro- or anti-feeling about a given proposition, whether one is experiencing the one kind of intuition or the other. For it does not seem easy to do so.

Looking through Maddy's (1998b) collection of "rules of thumb," one may guess that Gödel might consider some of them mere heuristic principles and others rational intuitions,[42] though ones too dim and misty to issue in rigorously formulated axioms rather than roughly formulated principles. But at least in my own case, I am able to guess this only because of my knowledge of attitudes Gödel has expressed in his writings, not because I myself have any sense when contemplating "rules of thumb" that *this* one is rational, *that* one heuristic.

Since Gödel does not discuss questions of heuristics explicitly, he never confronts the question of how to distinguish genuine, if fallible, perceptions of concepts from mere hunches perhaps based on subconscious inductive or analogical reasoning. But it seems there will be a serious weakness in his position unless a satisfactory answer can be given. At the very least, there will be a serious disanalogy between rational and sensory intuition. For when it comes to the visible properties of visible objects, there is no mistaking *seeing* what they are from having a *hunch* about what they must be.[43]

In the case of intuitions of concepts, unless there is a comparably unmistakable contrast between rational and heuristic intuition, the analogy between the former and vision will not be good. And while the length of

---

[41] The suspicion was confirmed by Cohen just a little too late for any more discussion than a very short note at the end to be incorporated into the paper.

[42] Especially the example Maddy calls "Maximize." This looks closely related to Gödel's thinking in footnote 23 (Gödel 1990, p. 262), already cited.

[43] This formulation may need a slight qualification. Suppose you are walking through a city you have never visited before, and are approaching a large public building, but are still a considerable distance away, and that the air is full of dust. Despite distance and dust, you are able to form some visual impression of the building. You are equally able to make conjectures about the appearance of the building by induction and analogy, taking into account the features of the lesser buildings you are passing, which you can see much better, and of large public buildings in other cities in the same country that you have recently visited under more favorable viewing conditions. Owing to the influence of expectation on perception, it is just barely possible, if the building is distant enough and the air dusty enough, to mistake such a conjecture for a visible impression, and think one is seeing what one is in fact only imagining must be there. But these are marginal cases.

the list of good analogies that can be discerned may not in itself matter much, in this particular instance a breakdown in the analogy would have a direct bearing on the question that does matter, the substantive question with which I began: In order to explain the Gödelian experience, do we really need to posit "rational intuition," or will some more familiar and less problematic type of intuition suffice for the explanation?

For if there is no criterion to distinguish rational from heuristic intuition, skeptics are likely to doubt that there is any such thing as rational intuition, over and above heuristic intuition, and to suggest that cases where axioms "force themselves upon us" are simply cases of very forceful heuristic intuition, perhaps based subconsciously on some very forceful analogical thinking. The search for new axioms for set theory, which Gödel urged on us, will then appear to skeptics rather as the law appeared to the greatest skeptic of all:

> Sometimes, the interests of society may require a rule of justice in a particular case; but may not determine any particular rule ... In that case, the slightest *analogies* are laid hold of, in order to prevent that indifference and ambiguity, which would be the source of perpetual dissention ... Many of the reasonings of lawyers are of this analogical nature, and depend on very slight connexions of the imagination. (Hume 1751, § III, part II, ¶ 10)[44]

The Humean picture is very different from the Platonic, on which judges, in order to decide cases where the letter of the law does not unambiguously imply a decision, should direct the inner gaze of the understanding to the contemplation of the Form of Justice. The Gödelian picture on which, in order to decide questions where currently accept axioms do not imply an answer, set theorists should direct the inner gaze of their mathematical intuition to contemplation of the Concept of Set, is threatened by a skeptical suggestion similar to Hume's suggestion about the law, that supposed mathematical intuition is no more than laying hold of slight analogies and connections of the imagination. The absence of much explicit discussion of heuristics in Gödel leaves me not knowing from which direction to expect a Gödelian defense against such threats.[45]

---

[44] In the version edited by J. Schneewind, the passage appears on p. 29. There is, of course, this difference from the situation described by Hume, that it is not so clear that the interests of society or even of mathematics demand a ruling on the status of the continuum hypothesis.

[45] Parsons, in correspondence, suggests that Gödel might emphasize that potential new axioms force themselves upon us *as flowing from the very concept of set*, something that is rather obviously not the case with his implausibility judgments, though it is equally obviously not the case with the "square axioms" Gödel was later to propose. The danger I see with emphasizing this feature, in order to distinguish rational from heuristic intuition, is that it may make it more difficult to distinguish rational from linguistic intuition.

Table 1 *Taxonomy of intuition*

I. *sensory*	(i)
A. *empirical*	(ii)
B. *pure*	(iii)
1. *spatial*	(iv)
2. *temporal*	(v)
II. *nonsensory*	(vi)
A. *rational*	(vii)
1. *mathematical*	(viii)
a. *geometric*	
b. *chronometric*	
c. *set-theoretic*	
d. *other mathematical*	
2. *other rational*	
B. *linguistic*	
C. *heuristic*	
D. *other nonsensory (if any)*	

Summary Table 1 lists the types of intuition distinguished in this chapter, numbering the unsubdivided types. The exegetical question considered was whether Gödel's implausibility judgments are supposed to be cases of (iv) or (vi) or (viii). My answer was (viii). The substantive question considered was whether the phenomenon Gödel would explain by appeal to (vi) could be explained by appeal to (vii) or (viii). My answer was a tentative "no" for (vii) and a tentative "yes" for (viii).

CHAPTER 3

Gödel on how to have your mathematics and know it too

Janet Folina

1 Introduction

Famous for his revolutionary work in logic and mathematics, Gödel's views *about* mathematics may seem untenable; as Parsons notes, some have even considered them scandalous (Parsons 1995a, pp. 44, 45). Metaphysically, Gödel appears to be a straightforward, if somewhat naive, realist, while epistemologically, his views are more complex and include inferences to the best explanation as well as appeals to mathematical intuition. Why might these views be untenable? Naive realism itself troubles many philosophers, but the addition of mathematical intuition – associated, as it is, with Kantian anti-realist views about mathematics – seems to take us to the edge of inconsistency. In short, Gödel seems to be a realist about mathematical objects, but he appeals to an epistemological framework associated with anti-realist views.

Why shouldn't a realist appeal to intuition? After all, if mathematics reflects an independently existing reality, and it can be known, then it is important to have an account of how so. The difficulty is that, *pace* Platonic "recollection," it is not very clear what the mathematical realist has in mind by "intuition." Simply invoking intuition, with no plausible explanation of its connection to truth (such as Kant gives), is not really an *account* of how we access mathematical reality. In other words, what grounds our confidence in the harmony between the independent reality and the purported intuition by which we know it? A related problem is that Gödel seems to refer to several types of intuition, including both

My thanks go to Juliette Kennedy for inviting me to participate in this volume. Though I am by no means a Gödel scholar, Juliette assured me that she sought to include some "fresh voices." I researched and wrote this paper in that spirit. Thanks go to Emily Carson, Geoff Gorham, Juliette Kennedy and Lisa Shabel for reading drafts of this paper, and also to the following "FIG" colleagues for feedback on a presentation of some of these ideas: Geoff Hellman, Charlie Geyer, Kim Logan, Shay Logan, and Catherine St Croix. Thanks also to an anonymous reader for constructive feedback.

"intuition that" certain sentences are true and "intuition of" certain mathematical objects (see Parsons 1980 and 1995a for more on this distinction). Additionally, Gödel's relationship to Kant is in need of clarification. He admired Kant's philosophy, regarding it as an important influence on the development of his views (Wang 1988, p. 17). Yet he rejected Kant's philosophy as strictly speaking false. (Leibniz and Husserl were other strong influences on his general philosophical views at different times; see Gödel 1961; Wang 1988, p. 19; Parsons 2010.) That is, Gödel was aware that a theory of mathematical existence defined in terms of human construction (or constructability), such as Kant's, does not square with a realist approach to mathematical objects. Nevertheless Gödel appeals to Kantian epistemology regularly, and to Kantian synthesis for one crucial explanation of his later concept of mathematical intuition.

What is Gödel's conception of mathematical intuition? If it is Kantian in any interesting sense, how can it be reconciled with his realism? After appealing to Kant's notion of synthesis in connection with mathematical intuition of sets, Gödel denies that this leads to Kant's account of mathematical existence.

> It by no means follows, however, that the data of this second kind, because they cannot be associated with actions of certain things upon our sense organs, are something purely subjective, as Kant asserted. Rather they, too, may represent an aspect of objective reality, but, as opposed to the sensations, their presence in us may be due to another kind of relationship between ourselves and reality. (Gödel 1947, reprinted in Benacerraf and Putnam 1983, p. 484)

Clearly Gödel saw his epistemological views about mathematics as aligned with Kant's, yet his metaphysical views as quite distinct.

Following Gödel, I will argue in this chapter that his realism is indeed consistent with his appeal to mathematical intuition. I will also argue that this consistency does not require a non Kantian, mysterious, occult form of intuition. The chapter divides into two sections. In Section 2, I defend Gödel from the charge of philosophical naiveté. I argue, on the contrary, that at least some of his arguments for realism (as well as intuition) take the form of "inference to the best explanation" (IBE), which is a common argument for scientific realism.

Section 3 focuses on Gödel's concept of intuition and the question of the extent to which it can be considered Kantian. My strategy, following Gödel, is to first detach the epistemic from the ontological roles of intuition in Kant. Next, I argue that at least one of Gödel's explanations of intuition is best thought of as quasi-Kantian. I then provide some historical context, that previous mathematicians endorsed similar, quasi-Kantian concepts of intuition helps to clarify Gödel's concept of intuition, and its position in the Kantian tradition. I conclude by considering two remaining obstacles to the thesis that Gödelian intuition and realism can coexist.

2 IBE arguments for realism and intuition

2.1 Gödel's mathematical realism

Gödel's strongest argument for realism about sets relies on an inference to the best explanation. For example, in "Russell's mathematical logic" (Gödel 1944; reprinted in Benacerraf and Putnam 1983, pp. 447–469), Gödel argues that the *fruitfulness* of mathematics indicates that it matches some genuine features of reality. The mainstream course of setting up "a consistent theory of classes and concepts as objectively existing entities" provides evidence for the objective existence of classes and concepts (Benacerraf and Putnam 1983, p. 468). Gödel further notes that the evidence here is of the same sort as in physics (Benacerraf and Putnam 1983, p. 456).

Of course there are differences. A limitation in set theory is that our concepts are not yet sufficiently clear, and they needed to be developed past the paradoxes. It is true that our human concepts *as manifested* at a certain time in mathematics are in part a social-historical contingency. But this does not mean that they fail to reflect real content.

> Classes and concepts may, however, also be conceived as real objects, namely classes as 'pluralities of things' or as structures consisting of a plurality of things and concepts as the properties and relations of things existing independently of our definitions and constructions. It seems to me that the assumption of such objects is quite as legitimate as the assumption of physical bodies and there is quite as much reason to believe in their existence. They are in the same sense necessary to obtain a satisfactory system of mathematics as physical bodies are necessary for a satisfactory theory of our sense perceptions... (Benacerraf and Putnam 1983, pp. 456–457)

The fruitfulness of set theory, and the agreement among mathematicians over the central axioms, provides evidence that the axioms successfully capture some facts.

There is a hint of Quinean postulation here, in that ontological commitments are determined by our best scientific theories. Though in many ways Gödel was not a Quinean (e.g., as Parsons points out, in his realism about concepts (Parsons 1995a, p. 53)), the argument above supports the view that ontology follows successful scientific theory. This seems rather sensible and not at all naive.

More specifically, Gödel believed that the success of set theory supports the soundness of its main concepts; and if its concepts are sufficiently clear then set theory must reflect some features of reality.

> ... if the meanings of the primitive terms of set theory ... are accepted as sound, it follows that the set-theoretical concepts and theorems describe some well-determined reality, in which Cantor's conjecture must be either true or false. Hence its undecidability from the axioms being assumed today can only mean that these axioms do not contain a complete description of that reality. (Benacerraf and Putnam 1983, p. 476)

Though "sound" concepts denote features of reality, the reality delimited may not be *completely* captured by our concepts. This is because conceptual boundaries may be vague or imperfectly understood at any given time.

We can compare this to the way natural kind concepts, for example on Kripke's "rigid designators" view, may capture some real kinds without the concepts themselves being quite correct, or fully, defined (see Kripke 1980). As with our former ostensive concept of "gold," Gödel believes the set-theoretic axioms succeed in designating the universe of sets while not precisely, or completely, describing this universe. That some truths are as yet undetermined does not mean, for either natural or mathematical kinds, that no truths have yet been discovered.

It is therefore clear that Gödel used IBE-style arguments to advance realism about sets. It is also clear that Gödel did this by explicitly mirroring similar arguments for natural science realism. Interestingly, Gödel uses a similar IBE-type of argument for mathematical intuition.

2.2   Gödel's arguments for intuition

Gödel added a supplement to the second edition of "What is Cantor's continuum problem?" (reprinted in Benacerraf and Putnam 1983, pp. 470–485). Most of his comments are brief; the fourth is a long and philosophically rich set of new remarks about mathematical intuition and its connection to set theory realism. Here it becomes clear that for Gödel, intuition supports, rather than competes with, the realist outlook on mathematics.

Gödel appeals to mathematical intuition to defend his view that the set-theoretic axioms are true or false, that is, they are not mere stipulations. Set theory may not (yet) give us a *complete* account of sets; but the set concept cannot be regarded as a convention.[1] For Gödel, set theory codifies fundamental features of our intellect (as well as reality); and like the intuitionists who opposed logicism about arithmetic, Gödel appeals to intuition to explain the centrality and cognitive depth of set theory.

First, Gödel argues that mathematical intuition, in the form of intuition of sets, must exist.

> But, despite their remoteness from sense experience, we do have something like a perception also of the objects of set theory, as is seen from the fact that the axioms force themselves upon us as being true. (Benacerraf and Putnam 1983, pp. 483–484)

Gödel proposes that we have something *like* sense perception of sets. But he does not appeal to his own *experience* of intuition, and the intuition in question is not the *feeling* of certainty. Instead, the argument is IBE in form. The set-theoretic axioms force themselves upon us, and something like a perception of sets is the best explanation for this fact.[2] That is, the axioms are about something, namely the objects of set theory; and we know the axioms are true in something like the way we know empirical truths. As with physical objects, the fact that we know inter-subjective truths about sets confirms their objective, independent existence.

Intuition, in other words, is what explains the certitude of the axioms. It is also the mathematical analog to sense perception, providing the link between knowledge (of the set-theoretic axioms) and reality (the set-theoretic universe). Intuition is the best explanation for the agreement and certainty about the basic set-theoretic axioms: how could the axioms force themselves upon us without something doing the forcing – without our having (intuitive) access to the set-theoretic universe that the axioms are about?

Does the IBE style of argument for intuition mean that Gödel's concept of intuition is not Kantian? It need not. For example, Poincaré was a Kantian about mathematical intuition, yet he argues thus:

> Why then does this judgment [reasoning by recurrence, or induction] force itself upon us with an irresistible evidence? It is only because it is the

---

[1] As Parsons points out (1995a, p. 67, note) this reverses an earlier suggestion that the different approaches to set theory might be regarded similarly to the different geometries.

[2] A good question is, which set-theoretic axioms and who perceives them? It is plausible to interpret Gödel as believing that there is a core set theory about which most mathematicians agree and these axioms are forced on us.

affirmation of the power of the mind which knows itself capable of conceiving the indefinite repetition of the same act when once this act is possible. The mind has a direct intuition of this power, and experience can only give occasion for using it and thereby becoming conscious of it. (Poincaré 1902, p. 39)

For Poincaré as for Gödel intuition is appealed to in part to explain how the truth of an axiom can be forced upon us.[3]

Gödel goes on to assert that it is a psychological fact that such an intuition exists, and he sees this not only as supporting the truth of the core axioms, but also as supporting realism.

> The mere psychological fact of the existence of an intuition which is sufficiently clear to produce the axioms of set theory and an open series of extensions of them suffices to give meaning to the question of the truth or falsity of propositions like Cantor's continuum hypothesis. (Benacerraf and Putnam 1983, pp. 484–485)

Set theory may be incomplete, in that it allows undecidable questions, but it is meaningful to consider whether an undecidable proposition is true or false. That is, the clarity and apparent truth of the axioms urges the view that set theory has a definite subject matter. And this definiteness (of subject matter) in turn bestows determinacy of meaning on (some) unprovable propositions. Since the axioms leave certain propositions undecidable, the determinacy of the subject matter, Gödel argues, requires both realism and intuition. Intuition thus yields axioms, which may be currently incomplete, but which are sufficiently definite to give us a sense of the reality of the domain so (far) defined. Criteria of truth and falsity are thus meaningfully applicable not only to known results but also to new axioms.

A circularity concern should perhaps be addressed before moving on. As noted, Gödel proposes that intuition supports realism about unproved propositions. But intuition is also supported *by* realism – by the fact that the truth of the axioms is forced upon us. The combination of these two views may appear to result in a circle. That the axioms are true shows that we have mathematical intuition of sets. And because we have mathematical intuition of sets we can justify the realist view that unproved (and unprovable) propositions can have definite truth-values.

The appearance of circularity is, however, resolved by appeal to the analogy with empirical truth in the context of similar IBE reasoning, which goes roughly as follows. The fact that there is a body of empirical truths

---

[3] Kant's argument in the *Prolegomena* (Kant 1783) can perhaps be seen similarly: a priori intuition is the best (only) explanation of arithmetic and geometric knowledge.

that is generally accepted by the scientific community is evidence that humans have a sensory apparatus that is both (i) shared and (ii) reliable, i.e., capable of providing insight into the way things are. This success story also gives us confidence in our ability to discover new truths. Gödel is here arguing similarly. Set theory (at least some core) is a body of axioms generally agreed upon by mathematicians; so it provides evidence that we have some apparatus – mathematical intuition – that is both (i) shared and (ii) capable of providing knowledge of mathematical reality. Furthermore, the success of set theory gives us confidence in the possibility of discovering new truths about sets, such as a new axiom that would decide the Continuum Hypothesis. The form of argument seems no more circular in the mathematical than in the empirical setting.[4]

3   Intuition: the concept and its context

3.1   Interpretative options

Gödel explicitly links his concept of intuition to Kant. But how Kantian is his epistemology of mathematics? The secondary literature is split. One option, which seems to be the mainstream view, is that he is not Kantian. Owing to his mathematical realism, Gödel's concept of intuition is different from Kant's. The other basic option is that while Gödel's mathematical realism differentiates his overall philosophy of mathematics from Kant's, some aspects of his epistemology constitute a genuine link to Kant.

Supporting the first option are arguments by several notable Gödel scholars, such as Parsons (1995a), Tait (2010) and Burgess (Chapter 2). On this interpretation, Gödel's overall philosophy entails a conception of intuition that contrasts with Kant's in that it is detached from sensibility, and which may be better understood in terms of a pre-Kantian tradition. After providing a detailed taxonomy of various types of intuition, Burgess rejects the view that Gödelian intuition – especially intuition related to set theory and the continuum – is a Kantian "sensible" intuition. This reading is based largely on Gödel's remarks about evidence for new mathematical axioms, and about his commitment to "concept realism." The latter, in particular, points to a view of mathematical intuition as providing direct insight into concepts, and thus to a kind of "intellectual" intuition, which Kant would have rejected. Burgess supports a reductionist view whereby

---

[4] However, in the conclusion I will raise a further question regarding the appeal to mathematical "success."

several of Gödel's apparently different appeals to intuition can all be understood as merely "heuristic" intuition.

Parsons and Tait also argue that we should distinguish Gödel's concept of mathematical intuition from Kant's on the grounds that for Gödel, mathematical intuition is not finite: it necessarily transcends the notion of finite construction in intuition that Brouwer and Hilbert, for example, developed for mathematics.[5] Tait argues that "concrete" intuition only justifies a certain kind of finitism about mathematics, whereas Gödel's intuition is far more "abstract." The abstract nature of Gödel's concept of intuition leads Parsons to conclude that it is closer to a theory of reason than to a Kantian theory of intuition (Parsons 1995a, pp. 45, 56–64).[6]

In favor of the other interpretive option, Hallett (2006) supports a closer connection between Kant and Gödel. Like Burgess, he finds that Gödel appeals to a "heuristic" notion of intuition when considering the question of new axioms. But he argues that Gödel also appeals to a concept of intuition that is closer to perception. Hallett agrees with Parsons that even this latter is more like an "intuition that" certain sentences are true rather than an "intuition of" objects (owing to its close connection to the *concept* of set). Yet Hallett pursues a further way in which Gödel's epistemology is Kantian, explicating a suggestive remark of Gödel's about synthesis, which links Kant's notion of synthesis with intuition of sets (remarked on, but not pursued, by both Parsons (1995a) and Wang (1988, pp. 205, 303)).

Following Hallett, I will attempt to strengthen the connection between the epistemological views of Gödel and Kant, emphasizing that Gödel links intuition to sensibility – not as spatiotemporality but in terms of other abstract elements in Kant's theory of the constitution of empirical objects. I will argue further that this places Gödel's epistemology of mathematics in a tradition of other "Kantian" mathematicians from the early twentieth century, such as Poincaré. I begin by detaching the epistemic from the ontological roles of intuition in Kant.

3.2 *Two roles for intuition in Kant*

For Kant, *a priori* intuition has two tasks. One is to frame experience, in the sense of providing experience with a form – that of spatiotemporality.

---

[5] This takes the Hilbert development of Kantian intuition as a paradigm, where mathematical intuition is exemplified by finite, or "concrete" construction.

[6] Parsons (1995a) also shows that Gödel's realism and his appeals to intuition develop along separate historical lines, with the explicit commitment to intuition occurring later.

Spatiotemporality for Kant thus has a metaphysical role, in that it aids in constituting, or shaping, the objects of experience. It also has a constitutive role in mathematics. A priori intuition is what determines the field of possible mathematical objects, or the ground on which mathematical construction depends. Because the ontology of mathematics is in this way determined by the form of experience, Kant's view has the added bonus of yielding the automatic applicability of all mathematical truths.[7] The important point here is that the constitutive role of mathematical intuition in Kant's philosophy is typically understood as implying a constructivist view of mathematical existence. Mathematical objects do not exist independently of the minds of mathematicians, for they are constructed *by* those minds, owing to the form of experience *of* those minds.

A priori intuition also has an epistemological role for Kant, for it is what enables us to know the synthetic truths of mathematics in an a priori way. Mathematical knowledge is not analytic, for it typically extends the content of the concepts involved in the statement known. Yet it is a priori because in so reaching beyond the concepts it nevertheless remains independent of empirical facts. "Construction of concepts" is the way Kant explains how the mathematician can help herself to this additional content in an a priori manner. Constructions, or instantiations, of concepts, link the objects they denote with their formal properties – those given by the form of experience. The knowledge thus gained is synthetic, since the concepts are supplemented; yet it is a priori, since the information added is given a priori by the form of experience.

We can elucidate this by distinguishing the question "what is true of the *concept triangle*" from the question "what is true of *triangles*." Concepts lie in the purview of the understanding, but triangles are objects *in space*. Because, Kant argues, mathematicians prove things about triangles, and not just the triangle-concept, they access the further properties that spatial triangles have, properties that are nevertheless a priori since space is nothing beyond the a priori form of experience.

Intuition thus has a crucial epistemic role for Kant in that it provides an explanation for mathematical methodology. Intuition not only determines which mathematical objects exist (as possible mathematical constructions); it also explains how mathematicians gain synthetic a priori knowledge about these objects.

---

[7] Indeed Rohloff (2010) has argued that explaining applicability was a central goal of Kant's philosophy of mathematics.

The main difference between realist and anti-realist appeals to intuition is that realist "intuition" does no fact-constitutive work; it is purely epistemic. Two problems then arise. The more minor problem is inter-pretive: how can one substantiate the view that in appealing to intuition, a realist such as Gödel is genuinely connected to the Kantian program? In other words, can we make sense of Kantian intuition without its metaphysical role? Kant appeals to intuition in order to explain the apparent amplification of concepts necessary for mathematical judgments, which he thought were obviously substantive and synthetic. Given Gödel's realism about both the objects and concepts of mathematics, he did not regard intuition as contributing additional substantive *content* to math-ematical concepts. And this seems an important difference between the two views.

The more substantive concern is a skeptical problem for any realist. That is, even if it is granted that we have mathematical intuition, how can the realist justify her view that this intuition yields insight into the independently existing mathematical reality? Why should we expect that our intuition is in harmony with reality, when the reality in question is of the independently existing sort? These are two of the difficulties to overcome in justifying Gödel's epistemology as Kantian.

3.3   Gödel's appeal to intuition

Let us first recall Gödel's analogy between intuition of sets and sense perception of empirical objects.

> But, despite their remoteness from sense experience, we do have something like a perception also of the objects of set theory, as is seen from the fact that the axioms force themselves upon us as being true. (Benacerraf and Putnam 1983, pp. 483–484)

This has the unfortunate connotation that mathematical intuition is akin to a mysterious "sixth sense," especially if the second half of the sentence is overlooked. Gödel does immediately qualify this first remark with a series of further comments about intuition. We will consider these in turn.

Gödel begins by acknowledging that mathematical intuition of sets does not appear to grant us any immediate knowledge of sets, which may seem to undermine the original proposed analogy between sense perception and mathematical intuition. However, as Gödel quickly reminds us, sense perception does not typically grant immediate knowledge of empirical objects either. Intuition of sets should not be expected to yield immediate knowledge of truths about sets any more than sense perception should be expected to yield immediate knowledge of empirical truths.

> It should be noted that mathematical intuition need not be conceived of as a faculty giving an *immediate* knowledge of the objects concerned. Rather it seems that, as in the case of physical experience, we *form* our ideas also of those objects on the basis of something else which *is* immediately given. (Benacerraf and Putnam 1983, p. 484)

For Kant, also, though a priori intuition is itself immediate (unmediated), knowledge via intuition is not, in general, immediate. Most arithmetic and geometric judgments are not immediate for Kant. This is why "construction of concepts" is necessary for mathematical knowledge: aside from a privileged few judgments (axioms), mathematical knowledge requires the active instantiation of concepts, by considering the *objects* they denote.

Gödel appeals to the active features of Kant's epistemology to explain further the role he sees for intuition in mathematics. The following, second, remark elaborates on the *formation* of ideas of objects.

> even our ideas referring to physical objects contain constituents qualitatively different from sensations or mere combinations of sensations, e.g., the idea of object itself... (Benacerraf and Putnam 1983, p. 484)

Ideas of objects do not result immediately from sensation. Rather they must be *formed*.

I wish to emphasize two things. First, Gödel stresses that formation of ideas requires a *process*. Second, Gödel endorses the view that such a process requires that there be some abstract and a priori contribution over and above the plurality of sensations. Gödel is correct that this perspective is essentially Kantian. Even granting that we can combine several sensations on a purely empirical basis,[8] this mere combination does not give us an object, a unity, but only a plurality.[9] To have an *experience of a table* requires that we unify certain sensations, exclude others, and further experience that there is a *thing* associated with those sensations. So Gödel agrees with Kant that in order to have ideas of objects, sensations must be synthesized, or grouped together, under certain concepts – especially a general concept such as "object."

How does this Kantian view about the role of synthesis in knowledge of empirical objects relate to Gödel's views about mathematics? In a

---

[8] This is actually not obvious to me, since we would have to hold them in our minds together – a holding which itself must involve an activity of the mind.

[9] Hallett (2006, p. 125) explains this aspect of Kant.

third remark Gödel asserts that the positing, or production, of the *idea of object* – in addition to our sensations – is "closely related to" the mathematical given.

> Evidently the 'given' underlying mathematics is closely related to the abstract elements contained in our empirical ideas. (Benacerraf and Putnam 1983, p. 484)

By the "abstract elements" I take it Gödel includes the idea of "object" referred to in the prior sentence; by the mathematical "given" I take it he means mathematical intuition, and in particular, mathematical intuition of sets.

The relation to sets is confirmed in a fourth remark about Kantian "synthesis," which follows the above immediately in a footnote.

> Note that there is a close relationship between the concept of set explained in footnote 14 and the categories of pure understanding in Kant's sense. Namely, the function of both is 'synthesis', i.e., the generating of unities out of manifolds (e.g., in Kant, of the idea of *one* object out of its various aspects). (Benacerraf and Putnam 1983, p. 484, footnote 26)

Gödel refers here to an earlier explanation of why he supports the iterative concept of set rather than what one might call the "totality" concept. The argument is that unlike the abstract logical, or "totality," concept of set, mathematical sets are typically "sets of integers, or of rational numbers... or of real numbers... or of functions of real numbers... etc."

> This concept of set, however, according to which a set is something obtainable from the integers (or some other well-defined objects) by iterated application of the operation 'set of', not something obtained by dividing the totality of existing things into two categories, has never led to any antinomy whatsoever... (Benacerraf and Putnam 1983, pp. 474–475)

Gödel's claim is that the iterative concept of set, in *particular*, bears a "close relationship" with Kant's categories. Both guide the synthesis of the relevant objects via abstract a priori concepts. We experience and cognize empirical objects as discrete, or unified, and falling under concepts; for Gödel, as for Kant, this requires syntheses of sensations via abstract concepts such as "unity" or "object." Similarly, sets are cognized by iterated applications of set-synthesizing operations via the iterative concept of set.

A concern may arise that Gödel claims too many "close relationships." One is the analogy between sensation and mathematical intuition/perception. Another is that the "'given' underlying mathematics is closely related to the abstract elements contained in our empirical ideas" (Benacerraf and Putnam 1983, p. 484). But, if intuition is the mathematical "given," then intuition has been compared both to sensation *and* to what is a priori *added to* sensation when forming ideas of empirical objects. Furthermore, Gödel claims a third relationship between the concept of set and Kant's categories, in that the function of both is to synthesize, or generate unities.

Is mathematical intuition analogous to sensation or to the Kantian categories, which must be *applied to* sensation? Or is it better compared to Kant's "imagination," a third classification of mental activity that unites the understanding with sensibility (via the schemata, or rules, that accompany concepts)?

Perhaps Gödel did not sharply differentiate among these options. Because he connected realism about objects to realism about concepts, it is unlikely that we will find a sharp distinction between *intuition* of sets as analogous to sensation, and the *idea* of set as an a priori category analogous to Kant's a priori categories. I will sketch the following interpretation in an effort to reconstruct the various analogies, and "close relationships."

(i)   Gödel endorses the Kantian view that we need *both* sensations *and* synthesis via concepts (with Gödel emphasizing the idea of object) in order to form ideas of empirical objects:

SENSATIONS + *synthesis* via the {idea of object} ⇒ ideas of objects.

(ii) Gödel next focuses on the abstract, formal part of Kant's framework – that which is a priori added to raw sensation (which for Gödel is something like *synthesis* via the {idea of object}) – and proposes that this, like sensation, produces knowledge of the "given" in mathematics. So the given in mathematics, which forms its basic content – for example, the set of integers – is accessed via the a priori formal element in experienced objects (*synthesis* via the {idea of object}). On this interpretation, mathematical intuition (the analog of sensation) results from the active processing required for empirical knowledge (that which is necessary to convert raw sensations into experiences and ideas of *objects*).

(iii) Gödel sees a further important similarity between *synthesis via the concept of object*, which yields ideas of empirical objects, and *synthesis via the concept of set*, which, applied to basic sets, yields (ideas of) the rest of iterative set theory. So for Gödel:

BASIC SETS (resulting
from abstract idea of   *synthesis* via the   iterative hierarchy
synthesis via the   +   {concept of set}   ⇒   of sets.
{idea of object})

The agreement with Kant is important, if schematic.[10] The formal element of experience, the abstract processing of sensation to form ideas of empirical objects, provides us with "mathematical intuition" of the basic mathematical objects (as well as some further operations). On this reading, rather than spatiotemporality, Gödel credits the operational, active contributions of the mind as providing access to the ontology of set theory, in its provision of access to basic sets and operations. Intuition is thus shifted from spatiotemporality to the more abstract tools of synthesis and the idea of object. It is a shift; yet for Gödel as for Kant mathematical intuition is a priori given via something that must be added to raw sensation in order to constitute (ideas of) objects.

The link to Kant may seem speculative, in its reliance on Gödel's use of the term "intuition" and the appeal to Kant's combinatorial notion of synthesis and other object-constitutive contributions of the mind. I will make two further comments to support this interpretation.

First, though the activity of synthesis via concepts is, in Kant's framework, rather different from spatiotemporal intuition, its centrality to Kant's account of mathematical knowledge should not be underestimated. For example, Emily Carson has argued that what is really fundamental for Kant's account of arithmetic knowledge is *synthesis* of intuition via the category of quantity, which takes place *in* time but which is distinct from the intuition *of* time (Carson 2011). Tait also emphasizes the role of synthesis in his explanation of Kant on mathematical knowledge (Tait 2010). In addition to the crucial role of synthesis, we can note that Kant himself emphasizes schemata over intuition in his explanation of "construction of concepts," which is central to his mathematical methodology (Kant 1783, 1787, B179–181). The centrality of both synthesis and schemata to Kant's views about mathematical epistemology suggests a more complex and cognitive story than one whose primary focus is on spatial/temporal intuition.

A legitimate worry remains, however: how can a Kantian conception of synthesis – which grounds the constitution of ordinary physical objects and takes place successively, in time – suffice for ideas of infinite sets?[11]

---

[10] And if ignoring, for now, the issue of spatiotemporality – in particular, the fact that synthesis takes place in time for Kant.

[11] This is one of the reasons for Parsons's negative assessment of Gödel as a Kantian.

To justify all the sets of classical set theory, Gödel's "synthesis" needs to be rather metaphorical; this, in turn, weakens the authenticity of Gödel's proposed tie to Kant. Perhaps, however, we can simply accept this difference as a consequence of his realism. That is, "synthesis" need not be ontologically limiting for a realist. Though our *experience of* empirical objects must be synthesized in time, neither empirical nor mathematical objects need to be so synthesized (for Gödel). Gödel argues only that synthesis forms the basis of our *understanding* of the set-theoretic universe; its constitution, its *existence*, is for him of course independent of this understanding.[12]

Second, as I will illustrate in the next section, post-Kantian constructivists had already shifted the term "intuition" towards Gödel's conception. Like Gödel, they connected intuition less to sensibility and more to various powers and activities of the mind – activities that, for Kant, may have belonged more to the purview of the imagination (the "home" of synthesis) than to sensibility (the "home" of intuition). Thus, by the late nineteenth to the early twentieth century, the term "intuition" had already evolved in the direction Gödel takes it. I will illustrate this evolution in the next section, comparing Gödel's views about intuition to those of several prominent mathematicians who preceded him.

There are probably many reasons for this change. Mathematics had become much more abstract and multifaceted, and the concept of mathematical truth had likewise evolved. In the light of these changes it was no longer plausible to consider mathematics as divided into the Euclidean/geometric science of space and the arithmetic/algebraic science of time (if it ever was). For a Kantian account of mathematics to remain coherent, the concept of intuition had to evolve too. It had to become more abstract and more flexible, in order to accommodate the more abstract, structural and diverse nature of mathematics.[13]

3.4   Intuition evolved: some historical context

Michael Friedman has divided recent attitudes towards the *a priori* into three classes (Friedman 2010). The first is Kant's conception of an

---

[12] I will return to this point in the conclusion.

[13] In some ways it reverted to a pre-Kantian, more cognitive, sense of intuition. (I owe this insight to Geoff Gorham.) This fits with recent assessments of Gödel's concept of intuition (e.g., Parsons 1995a).

absolute synthetic a priori framework, necessary for all humans to have the kinds of experiences and knowledge that we have. Second, there was a shift, especially among a certain subset of philosophically minded mathematicians, to modify Kant's concept of the a priori. Though some of the details of Kant's philosophy were eliminated, the basic idea that mathematics depends significantly on "built-in" features of the human mind was still attractive to people such as Poincaré, Helmholz, Brouwer, and Husserl. A third shift is to what Friedman calls the "relativized a priori," which he identifies with the role of convention in some of the logical positivists' work as well as in his own recent work in the philosophy of science and mathematics (see also Friedman 1999, 2001).

Seen through these categories, Gödel is best aligned with the second approach to the a priori. Although he was historically closer to logical positivism, and even attended meetings with some of its main proponents, it is well known that he disagreed with its core philosophical views. Both his realism and his appeal to mathematical intuition are at odds with positivism. And though his realism differentiates his philosophy from the mainstream "inheritors" of Kantian intuition, such as Brouwer and Poincaré, his concept of intuition is actually very similar.

Like Gödel, Poincaré strays from a strictly Kantian approach in several senses, while still considering himself a defender of Kant's philosophy of mathematics (in particular against the logicists and formalists). First, though he claims to be some sort of constructivist about mathematics, he wanted to avoid revising mathematics, as Brouwer ended up doing. The autonomy and freedom of mathematics was important to Poincaré. Thus, somewhat like Gödel, Poincaré wanted to have his mathematical cake and eat it too. For example, Poincaré occasionally speaks of the intuitive continuum as a source of mathematical content, or ontology – rather than as a manifold in which mathematical construction occurs – for more "geometric" areas of mathematics such as topology and analysis. Appeal to this intuition enables Poincaré to avoid strict constructivism about mathematical existence. (For example, he sometimes implies that we can simply *posit* the domain of real numbers, see Poincaré (1913), Chapter 3, especially p. 44.)

Second, mathematical intuition for Poincaré was not spatiotemporality. Though Brouwer appeals to time as the source of the intuitive continuum, as well as for the basic mathematical process of "two-oneness," Poincaré does not appeal to time in connection with intuition – at least not explicitly (compare Brouwer 1913, reprinted in Benacerraf and Putnam

1983, p. 80). The central mathematical intuition to which Poincaré appeals is that of "indefinite iteration" or repetition, explaining this intuition as a *power of the mind*, a mental capacity, rather than a form of sensibility.

> Why then does this judgment [reasoning by induction] force itself upon us with an irresistible evidence? It is only because it is the affirmation of the power of the mind which knows itself capable of conceiving the indefinite repetition of the same act when once this act is possible. The mind has a direct intuition of this power, and experience can only give occasion for using it and thereby becoming conscious of it. (Poincaré 1902, p. 39)

The kinds of acts Poincaré thinks we can envisage repeating endlessly are mental acts, such as *modus ponens* inferences. This enables us to recognize the truth of mathematical induction, for induction condenses an infinite sequence of *modus ponens* steps into a single sentence, codifying the leap "from the finite to the infinite" (Poincaré 1902, p. 38). Intuition is indispensable, for without induction mathematics would be significantly reduced. Thus, at the core of mathematical reasoning is a synthetic *a priori* judgment (Poincaré 1902, p. 39) which we are "forced" to accept as true because it merely "affirms" a power of which we have a direct intuition: the power to conceive of the indefinite repetition of an act.

This is similar to Gödel's appeals to intuition in the following three senses. First, Poincaré has appropriated the general idea of a priori intuition as something that explains the possibility of mathematical knowledge, detaching it from spatiotemporality in particular. Second, Poincaré connects this given more to a cognitive process than to sensation: the power of the mind to iterate certain operations. Third, as noted above (Section 2), Poincaré also deploys an IBE argument: intuition is offered as an explanation of the fact that the principle of induction is forced on us as true. The premise regards a truth – that of the principle of induction; the (best) explanation for our apprehending induction as indisputably true is the intuition of indefinite iteration.

Gödel's views on intuition are also comparable to those of the early Hermann Weyl. Weyl's predicative analysis is a mathematical embodiment of fairly strict philosophical/constructivist principles. With Brouwer, and unlike Poincaré, Weyl was willing (in his earlier work) to revise mathematics on philosophical grounds. Yet he shares with Poincaré the idea of mathematical intuition as neither space nor time but "iteration":

> Only when I had achieved certain general philosophical insights (which, incidentally, required that I renounce conventionalism),... I became firmly convinced (in agreement with Poincaré, whose philosophical position I share

in so few other respects) that the idea of iteration, i.e., of the sequence of the natural numbers, is an ultimate foundation of mathematical thought – in spite of Dedekind's "theory of chains" which seeks to give a logical foundation for definition and inference by complete induction without employing our intuition of the natural numbers. For if it is true that the basic concepts of set theory can be grasped only through this "pure" intuition, it is unnecessary and deceptive to turn around then and offer a set-theoretic foundation for the concept "natural number." (Weyl 1918, p. 48)

That certain mathematical truths are "forced on us" (Poincaré), or that the basic concepts in the axioms we take to be true require intuition (Weyl), both resemble Gödel's views about intuition. All depart from a strict sensibility-based conception. All blur the (purportedly) sharp Kantian distinction between sensibility and understanding. With Poincaré, Gödel appeals to intuition to explain how certain axioms can "force themselves on us" (Benacerraf and Putnam 1983, pp. 483–484). With Weyl, Gödel also emphasizes the set-theoretic *axioms* and the importance of *concepts* (rather than intuition of domains or objects) in our understanding the axioms as true (Benacerraf and Putnam 1983, p. 474).

Of course, Poincaré and Weyl were both drawn to constructivist views of mathematical existence while Gödel was decidedly not. However, in Frege we find a mathematical realist who also appealed to Kantian intuition (for geometry). Frege's modification of intuition departs from that of later intuitionists, who gave up the view that Euclidean geometry is (in particular) grounded in intuition, supporting instead the intuitive nature of arithmetic. In contrast, Frege supports the intuitive status of geometry, and argues that arithmetic does *not* require intuition. Both modifications of Kant result in a contrast, an asymmetry, between arithmetic and geometry. The point here is that for Frege the appeal to intuition is detached from a constructivist view of mathematical existence.

Frege contrasts geometry with arithmetic as part of his argument that arithmetic is logically independent of intuition. Even if we can consider the number 100,000 an intuition (because it is particular) it cannot ground general laws (Frege 1884, section 12). In contrast, geometric intuition *can* ground general principles.

> In geometry, therefore, it is quite intelligible that general propositions should be derived from intuition; the points or lines or planes which we intuite [sic] are not really particular at all, which is what enables them to stand as representatives of the whole of their kind. But with numbers it is different; each number has its own peculiarities. (Frege 1884, section 13)

For Frege, their *individuality* means that particular numbers cannot represent numbers in general. And intuition plays a genuine role in mathematics only when it can ground general principles, that is, when an instance can represent the whole class.

Frege's appeal to geometric intuition is intended to be Kantian. His argument is that unlike geometry, where Kant was right about intuition, Kant was wrong in thinking that arithmetic depends on intuition. But his concession to geometric intuition is *epistemological only*. Like Gödel, Frege is a mathematical realist: "the mathematician cannot create things at will, any more than the geographer can; he too can only discover what is there and give it a name" (Frege 1884, section 96). Frege's endorsement of Kantian intuition, and of the synthetic a priori character of geometric knowledge (Frege 1884, section 89), is not thereby an endorsement of Kantian constructivism about geometric objects. Mathematical intuition is determined by appeal to considerations of epistemology and methodology; intuition has no metaphysical or ontological consequences. Thus, it is in its *epistemic role only* that Frege defends geometric intuition.

For Gödel, as for Frege, intuition plays the epistemic role of grounding our knowledge of general principles – a role also endorsed by Poincaré and Weyl. Yet for Frege and Gödel, this role is detached from its fact-constitutive, ontological, role.[14] Around the early twentieth century, mathematicians who appealed to intuition tended to weaken its tie to sensibility. Even Brouwer, who is in some sense closer to Kant than Poincaré or Weyl, focuses less on features *of* time and more on what results from *thinking about* time in articulating his concept of mathematical intuition. The a priority of time may be the source of mathematical intuition, but the basic intuition of mathematics is for Brouwer the intuition of "two-oneness," not time. Two-oneness is a structural property, abstracted from our experience of time.

> This neo-intuitionism considers the falling apart of the moments of life into qualitatively different parts, to be reunited only while remaining separated by time, as the fundamental phenomenon of the human intellect, passing by abstracting from its emotional content into the fundamental phenomenon of mathematical thinking, the intuition of the bare two-oneness. (Brouwer 1913, reprinted in Benacerraf and Putnam 1983, p. 80)

We experience one thing and then another – experiences that are separated but joined through time. If we abstract from the content of those experiences we are left with the structure of one experience and (then) another;

---

[14] Dummett (1996) provides a helpful discussion of Frege's views about geometry, though Dummett is ambivalent about their coherence.

this structure is the "bare two-oneness" of which we have an intuition. The ability to form a new two-oneness from one of the elements of a given two-oneness (similar to the ability to form a new successor from a given successor) yields a *"process"* that can be "repeated indefinitely" and which in turn yields the smallest infinite ordinal number.[15]

So we see that despite Brouwer's more Kantian appeal to time as a source of intuition, time is not identical to mathematical intuition for Brouwer. Rather, mathematical intuition is something *produced from* the temporal nature of experience, and thus is more operational in the following ways. First, Brouwer's focus is on two-oneness, which is an abstract *structural* property of temporal experience. Second, this property is not really given *in* time; rather it has to be *formed* by thinking about experiences in time – hence its centrality to the "human intellect" rather than to the human sensibility. Third, Brouwer thinks of two-oneness as a *process*, which can then be indefinitely iterated, and this is the source of infinity. The focus on process and the connection to the human intellect in particular draw us away from the Kantian picture of intuition as the spatiotemporal form of sensibility.

As I see it, detaching the epistemic role of intuition from its ontological role, as both Frege and Gödel did, is simply another version of this general trend of post-Kantian constructivists to modify Kant's theory of a priori intuition in order to make it more compatible with the development of abstract mathematics. These are very programmatic remarks about the development of mathematical intuition after Kant, which will not be explored further here. What I hope to have shown is that Gödel's concept of mathematical intuition can be situated within this second historical trend of approaches to the a priori. It is thus arguably neither singularly naive nor particularly dissociated from Kantian epistemology.

4   Conclusion

I will close by addressing two final obstacles (mentioned above) to the combination of Gödel's quasi-Kantian epistemology with his realism. The first concern arises from a crucial difference in the relationships between concepts and mathematical intuition for Kant and Gödel.

For Kant, intuition is central to a complex account of how mathematicians supplement, or go beyond, concepts. Intuition contributes content

---

[15] Brouwer also appeals to the form of time as being continuous, and continuity also supplies important mathematical properties that shape Brouwerian analysis.

necessary for certain judgments via a process Kant calls "construction" of concepts. This seems utterly unlike Gödel's conception, where the distinction between intuition of sets and the concept of set, for example, is less definite. Intuition of sets is meant to explain our knowledge of the truths of set theory, but these truths seem more or less about the concept of set; set-judgments do not seem ampliative for Gödel as most mathematical judgments are for Kant. Rather, intuition, for Gödel, seems to provide insight into both concepts and the objects they denote, rather than into the properties of objects that do not strictly fall under (yet belong to) the relevant concepts. Thus Gödelian intuition is less clearly "synthetic"; it is closer to an analytic intuition.

However, owing to his realism about concepts, Gödel was beginning from an understanding of "concept" that was already quite different. Kant's concept of "concept" involved whatever sub-predicates a given concept contained. Gödel's realism means that concepts are associated instead with groupings of things. So a different relationship between intuition and concepts is somewhat inevitable.

In any case, as noted above, the general conceptual landscape had changed dramatically between the time of Kant and Gödel, particularly regarding the framework concepts, "logic," "concept," and "intuition." As we have seen, conceptions of mathematical intuition were already moving in the direction of formation of concepts rather than being strictly tied to sensibility. Thus, Gödel's association of intuition with both concepts and objects – though a departure from Kant – is nevertheless in the trajectory of late nineteenth and early twentieth century Kantian developments of mathematical intuition.[16]

The second obstacle is the skeptical problem: why should the appeal to intuition be granted to the realist? For Kant, the veracity of intuition is explained: intuition provides insight into mathematical truth because this same intuition is involved in *constituting* those truths. Intuition has justifying power (it is not merely a psychological/ causal mechanism) because Kant's link guarantees that mathematical beliefs are connected to truth. But for Gödel and Frege, who discard this fact-constitutive role, the question remains: what are our grounds for believing that intuition gives us insight into mathematics if the facts of mathematics are independent of us, including our intuition? That is, why should we think that reality conforms to our intuition?

---

[16] How Kantian are any of these developments? Again, in many ways they are closer to a pre-Kantian tradition (see Parsons (1995a); thanks again to Geoff Gorham for this point).

Why should we presume that psychology is in harmony with reality, if reality is independent of our psychology?

The scientific realist sometimes answers this question by appealing to success. If our scientific theories and senses were way off track then we/science would not be so successful. Is the same response open to the mathematical realist? Can Gödel or Frege call on the success of mathematics to justify the veracity of our mathematical intuitive apparatus?

It appears not, for there seems to be no independent check on mathematical "success." That is, unless we assume that successful mathematics yields *scientific* success/applicability, we have no independent measure of *mathematical* success. Gödel acknowledges the problem of skepticism for the realist, but he simply dismisses it. The issue arises just after he denies that certain aspects of Kant's epistemology – those he accepts – imply metaphysical constructivism about mathematics.

> It by no means follows, however, that the data of this second kind, because they cannot be associated with actions of certain things upon our sense organs, are something purely subjective, as Kant asserted. Rather they, too, may represent an aspect of objective reality, but, as opposed to the sensations, their presence in us may be due to another kind of relationship between ourselves and reality. (Benacerraf and Putnam 1983, p. 484)

Of course, mathematics was not subjective for Kant, as Gödel implies. Presumably, Gödel's purpose here is to reinforce the distinction between the two accounts of the objectivity of mathematics. Gödel's view of mathematical objectivity is stronger than Kant's. Though mathematics is not subjective for Kant, the objective validity of mathematics for Kant is delivered by the fact-constitutive role of a priori intuition rather than by the existence of an independent mathematical reality. Gödel distinguishes his metaphysical views about mathematics from the ontological, fact-constitutive, role of Kantian intuition, while asserting the similarity of their epistemic views.

It is here that Gödel admits that this combination is open to skepticism.

> However, the question of the objective existence of the objects of mathematical intuition (which, incidentally, is an exact replica of the question of the objective existence of the outer world) is not decisive for the problem under discussion here. (Benacerraf and Putnam 1983, p. 484)

Whether it is the physical world, or the mathematical world, realism *qua* realism allows the possibility of skepticism. That we have certain sensations, or intuitions, cannot guarantee that these faithfully reflect the independent, objectively existing entities. But, Gödel argues, just as skepticism is not fruitful in science, so it has no decisive bearing on the question of the existence or veracity of mathematical intuition. The dismissal strikes me as unsatisfying for, as noted above, unlike the empirical domain, in mathematics we seem to lack independent grounds for inferring success and thus dismissing skepticism. Perhaps more can be said on Gödel's behalf.

One might defend the veracity of intuition by linking it more explicitly to natural science, via views associated with structural realism. According to structural realism, reality is independent of our knowledge of it; if so, presumably the form of reality is independent of us as well. While we cannot claim to know reality directly, we can know it indirectly, via its mathematical, and otherwise structural/relational, form. One kind of evidence is the success of science; related to this is the fact that successful theories (on this argument) often bear structural/formal relationships to earlier theories. Finally, we evolved in this reality with this form, and so our intuition might be thought of as molded to the form of reality via evolution. That is, a Darwinian account can bolster the following vision, which roughly conforms to Poincaré's view (Folina 2010). It is not that reality is molded by intuition, as for Kant; rather, intuition is molded to reality via evolution.

Can this Darwinian, structural realist picture be extended to support mathematical realism as well as scientific realism? Rather than Kant's view that mathematics results from the form of *experience*, a realist view might be that mathematics results from the form of *reality*. Realism about mathematics could thus be connected to realism about the form of empirical reality. Reality has a certain form, which mathematics captures. And since we evolved in reality we evolved to have access also to its form via logic, concepts and intuition – though this form transcends us and is not of our own making.

This admittedly vague view is *consistent* with Gödel's views, and perhaps mildly *supported* by some of them. *Consistent* with it is Gödel's concept realism. Concept realism means that certain concepts reflect real categories, or general features, of reality. Intuition could in part mean we are psychologically inclined to conceptualize the world in certain ways. Why should this harmonize with reality? On a quasi-Darwinian account, mathematics, language, and conceptual frameworks might all be grounded in the facilitation of survival.

Supporting this picture is Gödel's conception of mathematical reality: it is not separate from empirical reality. He seemed to think there is just reality. As quoted above, mathematical data "...too, may represent an aspect of objective reality, but, as opposed to the sensations, their presence in us may be due to another kind of relationship between ourselves and reality" (Benacerraf and Putnam 1983, p. 484). There is just one reality: natural science captures some of its features while mathematics codifies others. Gödel proposes that mathematical intuition be understood as "another" way humans relate to reality, in addition to the ordinary senses. Indeed, Gödel hypothesized that we have some "physical organ" (Wang 1988, p. 190) analogous to our various senses, which accounts for our ability to form abstract concepts, such as those in mathematics. Perhaps Gödel thought that this concept-forming "organ" helps to distinguish the "real" abstract concepts from the illusory, just as our sensory organs do so for empirical concepts.

Finally, Gödel postulates that this physical organ is "closely related" to the neural "organs" that enable language. So his vision seems to be that our brains are adapted to pick out real abstract concepts, in a way that is similar to the brain's adaptation to language. The connection to the language "instinct" (Pinker 1994) supports the possibility of a Darwinian account of mathematical intuition. Finally, this makes our "intuition" and knowledge of mathematical reality not at all mysterious, since it relies on a Darwinian, physical explanation.

Does Gödel get to have his realist cake and know/eat it too? Does he offer a plausible epistemology consistent with his realism? On the one hand there are two further problems undermining the above Darwinian gloss on his realism. One is the fact that Gödel was skeptical that evolutionary processes were sufficient to produce human minds (Wang 1988, p. 197). And divine intervention seems to re-introduce the mystery of Platonic epistemology we had attempted to avoid. Second, even if a naturalistic account can make sense of mathematics as about the form, or other abstract features, of reality, it would have to be argued that these features are plausibly as rich as a realist such as Gödel would like.[17] On the other hand, I hope to have bolstered an interpretation of Gödel's philosophy of mathematics as no more, and possibly less, scandalous than other forms of mathematical realism which all – it seems to me – suffer similar, very interesting, difficulties.[18]

---

[17] For example, Ruse (1998) argues that a Darwinian framework can only justify the mathematics beneficial to our survival. Obviously, differentiating between finite numbers is advantageous, and these differences presumably reflect empirical differences. Four tigers are worse then one. It is much less clear how one could extend this picture to transfinite realms.

[18] Difficulties famously, and eloquently, pointed out by Benacerraf (1973).

PART II

The completeness theorem

CHAPTER 4

Completeness and the ends of axiomatization

Michael Detlefsen

1 Introduction

Gödel began his 1930 paper on the completeness of the functional calculus with the following description of Whitehead's and Russell's *Principia* project.

> Whitehead and Russell . . . constructed logic and mathematics by initially taking certain evident propositions (*evidente Sätze*) as axioms and deriving the theorems (*die Sätze*) of logic and mathematics from these by means of some precisely formulated principles of inference in a purely formal way (that is, [w]ithout making further use of the meaning of the symbols). . . .

> when such a procedure is followed the question naturally arises at once (*erhebt sich natürlich sofort*) whether the initially postulated system of axioms and principles of inference is complete (*vollständig*), that is whether it actually suffices to deduce (*deduzieren*) *every* logico-mathematical proposition (*jeden logisch-mathematischen Satz*), or whether, perhaps, it is conceivable that there are true (*wahre*) propositions . . . that cannot be derived in the system under consideration. (Gödel 1930c, p. 103 emphasis in the text[1])

Following this description of the *Principia* project, Gödel noted Bernays's earlier proof of an affirmative answer for the (logical) truths of the propositional calculus, and he proceeded to prove the same for the (logical) truths of the restricted functional calculus.[2] He began his 1931 paper on the incompleteness of the *Principia* system PM with a different description of the *Principia* project.

> The development of mathematics toward greater precision has led . . . to the formalization of large tracts of it . . . The most comprehensive formal

---

[1] For the most part I follow the translations in Gödel (1986).

[2] Gödel made a similar remark in Gödel (1930a). There, however, instead of speaking of the provability of every logico-mathematical proposition or every true (*wahre*) logico-mathematical proposition as the condition of completeness, the term he used was "correct proposition" (*richtigen Satz*), cf. p. 125. He seems to have seen these as saying the same thing, however.

systems that have been set up hitherto are the system of *Principia Mathematica* (*PM*) on the one hand and the Zermelo–Fraenkel axiom system of set theory ... on the other. These two systems are so comprehensive (*weit*) that in them all methods of proof today used in mathematics (*alle heute in der Mathematik angewendeten Beweismethoden*) are formalized ... One might therefore conjecture that these axioms and rules of inference are sufficient to decide *any* mathematical question that can at all be formally expressed in these systems. (Gödel 1931b, p. 145)

What interests me most in these remarks are (i) Gödel's descriptions of the procedure he took Whitehead and Russell to have followed in constructing the *Principia* system (the formalized version of which he called PM), (ii) his description of the purposes he took them to have been pursuing in so constructing their system and (iii) his description of the completeness ideal he believed to be naturally associated with systems constructed in this way and for these purposes.

There are problematic elements in (i)–(iii) and they indicate serious differences between Gödel's conception of the ideals of the *Principia* project and Whitehead's and Russell's own conception of them. It is with these differences that I will mainly be concerned here.

2 Gödel's ideals of completeness

The notions of completeness featured by Gödel were all one or another form of completeness with respect to the truth. The immediate form was completeness with respect to truths expressible in the language of the theory. The type of truth featured, however, varied from case to case. Sometimes the aim was to capture all the logical truths formulable in a given language or type of language. In other cases (e.g., theories purporting to formalize arithmetic or some version of set theory), it was to prove every truth pertaining to some more restricted subject matter.

Sometimes such differences in subject matter imply differences in the basic forms that completeness conditions may be expected to take. Completeness with respect to logical truths, for example, cannot be expected to satisfy a general law of excluded third according to which for every sentence in a given language, either that sentence or its denial will be true.

Completeness with respect to other types of truth (e.g., classically conceived truth for arithmetic or set theory), however, can reasonably be expected to satisfy such a principle. In such cases, to provide for the proof of all truths pertaining to a given subject matter amounts to adopting the following condition as an ideal.

Gödel completeness: Let T be an axiomatic theory whose language is $\mathcal{L}$. T is Gödel-complete in case for every sentence σ of $\mathcal{L}$, either σ is a theorem of T or $\neg\sigma$ is a theorem of T.[3]

Gödel completeness, I claim, is very different from the type of completeness that primarily motivated Whitehead and Russell in the *Principia* project. The completeness with which they were principally concerned is what I will call *descriptive completeness*.

Roughly speaking, descriptive completeness is completeness with respect to a *data* set of some type. In the case of the *Principia* project, this was the set of *accepted* propositions of established mathematical practice.

This is a very different type of completeness from Gödel completeness. It reflects the distinctive aims of what I will call *descriptive axiomatization*. These aims center on the idea of a formal axiomatic system's serving as a *formalization of* a pre-axiomatically given body of informal mathematical practice – in particular, a formalization of the theorems and proofs with which that body of practice is chiefly associated.

Whether a descriptive axiomatization satisfies such a condition is not much affected by whether it is Gödel-complete. This at any rate is what I will argue. To present this argument effectively, and to indicate its depth, it will be useful to have a little historical background to draw upon. It is thus to a sketch of this background that I now turn.

3 Descriptive axiomatization

The aim of descriptive axiomatization has generally been to bring logical organization to a presumed body of "facts" or data that is taken to want

---

[3] It is perhaps worth noting that Gödel was not the first to formulate such a notion of completeness. Hilbert formulated it in his 1928 Bologna address, published as Hilbert (1930b) (cf. Problem III, pp. 6–7). He saw it as a finitarily admissible substitute for the finitarily inadmissible notion of categoricity. Categoricity had in fact commonly been referred to as *completeness* in the foundational literature of the early twentieth century. Even before the publication of Hilbert's paper, however, Langford formulated a condition of what was Gödel-completeness as an appropriate condition to place on at least certain axiomatic systems (cf. Langford 1927, pp. 96–97). Finsler had also defined such a notion and referred to it as *formal decidability* (cf. Finsler 1926, p. 680).

In Gödel (1930b, 1930c, 1931b), the term Gödel generally used for this notion of completeness was "Entscheidungsdefinitheit." He used the term "vollständig" in Gödel (1931a), where he wrote:

> A formal system is said to be complete (*vollständig*) if every proposition expressible by means of its symbols is formally decidable from the axioms, that is, if for each such proposition A there exists a finite chain of inferences, proceeding according to the rules of the logical calculus, that begins with some of the axioms and ends with the proposition A or the proposition not-A. (Gödel 1931a, p. 203)

such organization. To achieve such organization, the thinking has been, an axiom system must be *descriptively complete* (in the terminology used here), or complete with respect to the data it is intended to organize. This was the type of aim pursued by Whitehead and Russell and also by the great majority of mathematicians of the nineteenth and twentieth centuries who pursued axiomatizations of more ordinary, *local* areas of mathematical practice (e.g., various areas of geometry, algebra, arithmetic, analysis, etc.).

For Whitehead and Russell, the "data" to be organized formed a very broad and comprehensive class of propositions, namely, the totality of all generally accepted propositions of ordinary mathematical practice as a whole. They saw the deductive capturing of (suitable translations of) this data as a premier constraint on the success of their project.

> In constructing a deductive system such as that contained in the present work, there are two opposite tasks which have to be concurrently performed. On the one hand, we have to analyse existing mathematics, with a view to discovering what premisses are employed ... and whether they are capable of reduction to more fundamental premisses. On the other hand, when we have decided upon our premisses, we have to build up again as much as may seem necessary of the data previously analysed ... *[T]he chief reason in favour of any theory on the principles of mathematics must always lie in the fact that the theory in question enables us to deduce ordinary mathematics.* (Whitehead and Russell 1910, volume I, preface, p. v, emphasis added)

As Whitehead and Russell saw it, then, the form of completeness of primary interest with regard to a system of *principles* is a type of descriptive completeness which requires that every commonly accepted proposition of ordinary mathematics (or, more accurately, an acceptable translation of every such proposition) should be deducible within it.

Given the foundational nature of their project, though, there was another basic completeness ideal as well, what I will call *analytic completeness*. The aim of the *Principia* project was to find the principles of existing mathematics. An important programmatic concern was thus to carry the analysis of the data through to such a point that it became clear that there were no more basic premisses to which the data could reasonably be traced.

It was thus important to know whether the analytic search for principles was as thorough as it might reasonably be – whether, in other words, it was *analytically complete*, or such as to assure rationally that no continuation of it would yield principles more basic than those identified.[4]

---

[4] "[W]e have to analyse existing mathematics, with a view to discovering what premisses are employed ... and whether they are capable of reduction to more fundamental premisses" (Whitehead and Russell 1910, volume I, preface, p. v).

Descriptive completeness and analytical completeness were thus the two principal completeness ideals of Whitehead's and Russell's foundational project. What I referred to above as more *locally* (as contrasted with foundationally) oriented descriptive axiomatization had a different aim. It was not intended to organize accepted mathematics as a whole, but only one or another subjectivally restricted part of it.

Huntington gave a basic characterization of descriptive axiomatization as it pertained to systems of this more local type. Its salient characteristic, in his view, was that it deductively organized a body of propositions, and so provided for the transformation of a body of "miscellaneous" or unorganized facts into a proper science concerning them.

> [A] miscellaneous collection of facts ... does not constitute a *science*. In order to reduce it to a science, the first step is to do what *Euclid* did in geometry, namely, to *select a small number of the given facts as axioms, and then to show that all other facts can be deduced from these axioms by the methods of formal logic.* (Huntington 1911, p. 158, emphasis in text)

Huntington's remark readily generalized and, so generalized, it became a common view of the aims and purposes of local axiomatization in the nineteenth and twentieth centuries.[5]

The central tenets of this view are the following.

1. Axiomatization generally takes place against the background of a *data set*.

2L. The elements of this set are the commonly accepted sentences pertaining to a given subject area.

3. The basic purpose of axiomatization is to organize a data set deductively.

4. To accomplish 3 fully, the axioms of a proposed axiomatization must be *descriptively complete* – that is, all elements of the data set must be deducible from the axioms.

Both varieties of descriptive axiomatization distinguished here (i.e., the local and foundational varieties) are intended to satisfy conditions 1, 3 and 4. These conditions are thus what I would count as the core conditions of adequacy on descriptive axiomatization.

Condition 2L, on the other hand, was generally conceived only as a condition on local axiomatization. For foundational axiomatization, it was

---

[5] See Boole (1954, ch. 1, p. 5), Veblen (1903, p. 309), Veblen (1904, pp. 343–344), Hilbert and Ackermann (1928, pp. 42–43), Reidemeister (1930, p. 62), Beth (1959, p. 139) and Kneebone (1963, pp. 201–202) for similar statements from a variety of different types of sources.

replaced by a condition embodying a broader understanding of what the data are. Specifically, it was replaced by a more encompassing view of the data, something along the following lines:

2F.   The data set for a foundational axiomatization is the union of the data sets of all local areas of mathematics.

One way in which local and foundational descriptive axiomatizations have generally differed, then, is in the perceived breadth of their respective data sets.

They may and have differed in other ways too. On at least some variants (e.g., that described by Huntington) the axioms of a local axiomatization are taken to *belong to* the data for which organization is sought. Specifically, they were taken to be "*given* facts" (Huntington's words, my emphasis).

In foundational axiomatization, on the other hand, this seems generally not to have been the case. Principles or axioms have not generally been taken to be part of the data they are intended to found. Rather, they have been taken to be obtainable only through the application of relatively demanding processes of analysis or regressive reasoning *from* the data, taken as truths that are in some appropriate sense *given*.

Foundational axiomatization has thus generally made use of a division of truths into levels, and its project has been to trace the truths that are given back to the deepest level of truth to which they may properly be traced.[6]

Such separation of the axioms from the data was clear in the *Principia*, where the axioms were taken to serve as the *principles* of the data.

These principles were also believed to be regressively (i.e., broadly inductively) *supported by* the data. For this reason too, then, they were not taken to belong to the data.

Russell emphasized this regressive-justificative aspect of foundational axiomatizations repeatedly. The following remark is typical.

> [I]n mathematics, except in the earliest parts, the propositions from which a given proposition is deduced generally give the reason why we believe the given proposition. But in dealing with the principles of mathematics, this relation is reversed. Our propositions are too simple to be easy, and thus their consequences are generally easier than they are. Hence we tend to believe the

---

[6] It should perhaps be noted here that even in those conceptions of local axiomatization where it is not assumed that the axioms belong to the data set, there is generally no presumption that the axioms are to be in any significant sense *deeper* than the data they imply. Rather, it is assumed only that the axioms can be used to organize the data set logically, not that they "cause," or explain or account for it.

premises because we can see that their consequences are true, instead of believing the consequences because we know the premises to be true. But the inferring of premises from consequences is the essence of induction; thus the method in investigating the principles of mathematics is really an inductive method, and is substantially the same as the method of discovering general laws in any other science. (Russell 1907, pp. 273–274)

In Russell's view, then, a foundational axiomatization represented a justificative scheme with a characteristic structure – data at the justificative base, principles ascending inductively from these, and theorems descending deductively from the principles.

Part of the basic structure of a foundational axiomatization conceived on the *Principia* plan was thus an inductive justificative linkage of the data of a foundational scheme to its principles.

In order to capture the salient differences between local and foundational axiomatizations, we must therefore not only replace 2L with 2F, but also add the following to the basic conditions of foundational axiomatization.

5F.   The sentences in the data set must be evident.
6F.   The evidentness of these sentences must, at least to a certain extent, be inductively transferrable to the axioms.
7F.   Unlike the axioms of a local axiomatization, which were generally conceived to belong to the data set, the axioms of a foundational axiomatization are not generally assumed to belong to the data set.

There are thus both local and foundational varieties of descriptive axiomatization. This notwithstanding, the varieties are unifiable under a single abstract description which Heyting made explicit in the 1960s.[7]

The pivotal element of Heyting's characterization is a distinction between a *pre-axiomatic* system of sentences $\mathfrak{D}$ (for data) and an axiomatic theory $\mathfrak{A}$ which is supposed to serve as an organization of it. As Heyting put it:

> In many cases an axiomatic theory $\mathfrak{A}$ is constructed in connection with a theory $\mathfrak{D}$ which existed before the axiomatization... One may then ask whether $\mathfrak{A}$ is sufficient to derive all the theorems of $\mathfrak{D}$; if this is the case, $\mathfrak{A}$ is said to be complete with respect to $\mathfrak{D}$. (Heyting 1963, p. 6)[8]

This is a unifying characterization of the notion of a descriptive axiomatization in that it applies to both local and foundational descriptive axiomatizations.

---

[7] The basic ideas were of course around in fairly definite form well before that (cf. Veblen 1903, 1904; Hilbert and Ackermann 1928).

[8] For reasons of English mnemonics, I have substituted the letter "$\mathfrak{A}$" for the letter "$\mathfrak{G}$" and the letter "$\mathfrak{D}$" for the letter "$\mathfrak{T}$" in Heyting's formulation.

Specifically, it covers both axiomatizatons of $\mathfrak{D}$ where the axioms of $\mathfrak{A}$ are chosen from the elements of $\mathfrak{D}$, and axiomatizations of $\mathfrak{D}$ where they are not.

Heyting's distinction between a pre-axiomatic set of theorems $\mathfrak{D}$ and an axiomatic system which axiomatizes $\mathfrak{D}$ thus suggests a general characterization of *descriptive* axiomatization. If we let $\mathfrak{A}$ signify an axiomatic system, then we can say that $\mathfrak{A}$ is a descriptive axiomatization of $\mathfrak{D}$ in the case that $\mathfrak{A}$ is complete with respect to $\mathfrak{D}$.

Use of the term "descriptive" in this connection is suggested by the fact that $\mathfrak{A}$ is required to capture something that is in some sense *given prior to* it. That a theory T should capture a pre-theoretic given d would seemingly be what is required for T to be *descriptively adequate* with respect to d.

4 Descriptive axiomatization and the classical plan of scientific investigation

The origins of the descriptive conception of axiomatization are ancient. Its core elements were in fact the core elements of the classical plan of scientific investigation.

The classical plan of scientific investigation

Commencement: Data (*protasēs*) are in some sense and in some respect "given" (e.g., by their being supposed, suggested, observed to occur, evident, etc.).

Regression (analysis): Methods of analysis are applied to the data in order to discern their reasons or causes (*apodosēs*).

Progression (synthesis): When these reasons or principles are identified, the direction of the investigation is reversed and the principles are used to demonstrate rigorously the data.[9]

Dialectic or the general art of rational investigation was thus commonly taken to comprise both methods of discovery (*artis inveniendi*), which is

---

[9] The regressive/progressive terminology follows Kant.

> **Analytic and Synthetic Method.** The *analytic* method is opposed to the *synthetic* method. The former begins with the conditioned and with that which is grounded and goes on to principles (*a principiatis ad principia*); the latter goes from principles to consequents, or from the simple to the composite. As the first could be called the *regressive* method, so the latter could be called the *progressive* method. (Kant 1992, §117.3)

what the so-called methods of analysis were generally taken to be, and methods of adjudication (*artis iudicandi*), which is what synthetic methods were generally taken to be.

Cicero (106 BCE–43 BCE) in fact maintained that every proper method of investigation proceeds according to some division of methods similar to that which the classical plan marks with its division between regressive and progressive methods.

> [E]very careful method of arguing has two divisions – one of discovering, one of deciding. (Cicero, p. 459)

Over time, this broad division of methods became an enduring part of the standard conception of scientific method. The following description from the *Port Royal Logic* was typical in this regard.

> Generally speaking, method may be called the art of arranging a series of thoughts properly, either for discovering the truth when we do not know it, or for proving to others what we already know.

> Hence there are two kinds of method, one for discovering the truth, which is known as *analysis*, or the *method of resolution*, and which can also be called the *method of discovery*. The other is for making the truth understood by others once it is found. This is known as *synthesis*, or the *method of composition*, and can also be called the *method of instruction*. (Arnauld and Nicole 1964, p. 302)

To simplify a bit, we may say that the movement of the classical plan is from (a) propositions which are given as evident, to (b) principles which in some sense underlie or account for these, thence back, by more rigorous deduction, to (c) the originally given propositions (or one or another type of clarified variant of them).

The plan of the *Principia* recapitulates this classical plan. It begins with theorems and not, as Gödel suggested, with axioms taken themselves to be originally evident. These initial theorems are regarded as data in the sense of being sentences that the eventually proffered principles must deductively capture if they are to be descriptively adequate.[10]

---

[10] Heyting made remarks which suggest that he thought there is more to descriptive axiomatization in the case of foundational theories than mere capture of certain "data" statements. In particular, he suggested that to be properly regarded as descriptive, a set theory must be intended to capture a pre-theoretically given concept of set. Since, however, in his view, the naïve conception of set was inconsistent, set theory could no longer plausibly be regarded as intended to capture a pre-theoretic concept of set. This being so, he suggested, axiomatic set theory cannot plausibly be regarded as a case of genuinely descriptive axiomatization. In his words:

From these data there is then a generally regressive or analytic search for principles. Once these principles are identified there is then, finally, a rigorous deduction of the original data from them. Such was the classical plan, and such too was the plan of the *Principia*. What set the *Principia* project apart from the more usual implementations of the classical plan was the intended comprehensiveness of its data set. It was intended to include not only the recognized theorems of a particular area of mathematics but the acknowledged truths of mathematics as a whole.

5   Descriptive axiomatization, descriptive completeness and Gödel completeness

Gödel's description of the procedure of the *Principia* as one which begins with evident axioms thus seems inaccurate and misleading. In particular, it does not do justice to what we have just described as the overall descriptive character of the *Principia* axiomatization with its presumption of a body of *data* and its regressive linkage of these to a set of principles eventually identified as their foundation.

His claim that the question of Gödel completeness "naturally arises at once" (Gödel 1930c, p. 103) for the principles of the *Principia* is thus misleading at best. It presents Gödel completeness as a question which ought rationally to arise at once for the *Principia* project and like foundational projects. In truth, though, Gödel completeness is a distinctly secondary concern for such projects – that is, for descriptive foundational axiomatizations – if indeed it is a concern at all.

The same can be said of the many descriptive axiomatic projects of the twentieth century. If I am not mistaken, then, (I) the *Principia* project, as well as most of the local axiomatizations of twentieth century mathematics,

---

In all other [i.e., non-set-theoretical] cases a model for the axioms was known beforehand, in some cases, as for instance in Euclidean geometry or in the arithmetic of the natural numbers, by pre-mathematical knowledge, in others, like in algebra, by purely mathematical structures. But pre-axiomatical set theory had proved contradictory; no reliable knowledge about general abstract set theory was available before the axiomatization. Thus the question of the existence of a model for the axioms is essential and it must be so. If set theory is not completely void, something must exist that fulfills the axioms, but this existence is only revealed by the axioms. (Heyting 1966, pp. 238–239, square bracketed phrase added)

I find this a curious remark. In a volume dedicated to A. A. Fraenkel, one might have expected some acknowledgement of the so-called iterative conception of set and the possibility that *it* might serve as a coherent pre-mathematical conception of set that an axiomatic set theory might reasonably be intended to capture. Apparently, though, and in my view curiously, Heyting did not see the iterative conception of set as a compelling pre-mathematical general conception of set.

are descriptive axiomatizations. For such projects, (II) the most compelling completeness ideal is descriptive completeness or completeness with respect to a pre-axiomatically given data set. Finally, (III) generally speaking, Gödel completeness neither implies nor is implied by descriptive completeness and it generally has no comparably compelling standing as an ideal of descriptive axiomatic projects.

The argument for (I) is essentially that which was sketched above in Section 3. I will now briefly indicate the arguments for (II) and (III).

The chief reason in favor of (II) hearkens back to the basic purpose of descriptive axiomatization – namely, the explicit logical organization of a body of accepted propositions not previously so organized.

The desirability of such organization was generally acknowledged by nineteenth and twentieth century foundational thinkers. The mathematician-philosopher Auguste Comte, in fact, described explicit logical organization as a core motive both of mathematical science and of science in general.

> [T]he spirit of mathematics consists in always regarding all the quantities which any phenomenon can present, as connected and interwoven with one another, with the view of deducing them from one another . . .

> Indeed, every true science has for its object the determination of certain phenomena by means of others . . . Every science consists in the co-ordination of facts; if the different observations were entirely isolated, there would be no science. . . science is essentially destined to dispense, so far as the different phenomena permit it, with all direct observation, by enabling us to deduce from the smallest possible number of immediate data the greatest possible number of results. . . Mathematics . . . pushes this to the highest possible degree . . . (Comte 1892, Troisième Leçon, pp. 37–38)

J. D. Gergonne made a similar point. Indeed, in his view, there could be and were cases where improving the internal organization of existing knowledge was a rationally more compelling goal of scientific development than the addition of new knowledge.

> [A] science perhaps recommends itself less by the multitude of propositions which make it up than by the manner in which these propositions are related and connected to one another. In each science there are certain elevated points of view by standing on which one can embrace a great number of truths at a glance. Viewed from a less commanding position, one could believe these truths were independent of one another. Seen from the elevated view, however, one realizes they stem from a common principle,

often even incomparably easier to establish than the particular truths of
which it is the abridged expression. (Gergonne 1826–1827, pp. 214–215)[11]

In both Comte's and Gergonne's views, then, the overall epistemic value of
organizing a data set can sometimes be greater than that of expanding it
through what Comte termed "direct observation."[12]

This should not surprise us, I think. Our situation as human knowers is
essentially this: certain propositions (the *data*) are "given" to us as evident
and, generally speaking, our capacity to expand the class of propositions
given in this way fruitfully and prudently is quite limited. It is therefore
important for us to develop our knowledge by means that do not require
proportional extension of that which is "given." Specifically, it is important
for us to develop the "data" at our disposal logically or inferentially and to
do this as fully as we responsibly may.

Viewed in such a framework, descriptive incompleteness would represent a type of developmental deficiency, namely, a failure to integrate all
elements of a data set into an axiomatization which is intended to serve as
an organizational scheme for it.

This, then, is essentially the argument for (II).

Support for (III) comes mainly from recognizing that Gödel
incompleteness does not generally impede realization of descriptive
completeness, which is a central goal of descriptive axiomatization.

Perhaps the deeper question, though, is why descriptive completeness
ought to be seen as a central goal of descriptive axiomatization. The
answer, at one level, is the simple and direct one given above – namely,
"To organize the *data* as comprehensively as is possible."

Pressed for a deeper answer – that is, for an explanation of why,
epistemically speaking, full implication of the data ought to be an ideal
of descriptive axiomatic organization – one might turn in either of two
directions.

One direction would be towards a presumed condition of adequacy for
mathematical theories – namely, that they should be adequate to the
accepted facts or data concerning the subject(s) they are intended to
organize.

---

[11] See Courant and Robbins (1981, pp. 214–215) for a similar statement from the twentieth century,
albeit one which emphasizes the importance of the axioms being few in number and individually
simple.

[12] I should perhaps add that there is little indication in either Comte or Gergonne that they opposed
the idea that the basic form of epistemic extension or gain is that of adding to the *extent* of one's
knowledge. Their central point seems rather to have been that addition to "direct observations" is
not the only form, or the only important form, such extension may take.

The other direction would be towards the type of maximal-utilization-of-evidensory-resources consideration indicated earlier. The more fully a set of axioms captures the *data*, the more fully it utilizes their regressive justificative potential.

This at any rate would appear to be so if we suppose that (i) the *data* comprise the pre-axiomatically evident truths concerning the subject(s) being axiomatized and that (ii) the evidentness of the data (or some part of it) is transferrable to the axioms by available means of regression from the data to them.

For convenience, I will call theories that are complete with respect to their data in this way *evidentially complete* theories. What is crucial for evidential completeness in the sense I have in mind is that those propositions which are evidentially given should be regressively utilized to the fullest rational extent. Evidential completeness as I am thinking of it thus represents a type of fullness or completeness of utilization of a limited resource.

Evidential completeness would appear to imply that all propositions that are given as evident should at least ideally be used to improve the regressive support for a set of axioms that is supposed to serve as a means of logically organizing them. If the data for a proposed axiomatization are taken to consist of the evident propositions that pertain to its subject, then its descriptive completeness should imply its evidential completeness.[13]

This being so, descriptive completeness would appear to maximize the strength of regressive support for the axioms of a descriptive axiomatization. Moreover, so far as I can see, Gödel completeness offers no prospect of improving on this. The regressive justification of a set of axioms is in proportion to the extent of their known evident consequences. That a set of axioms should be Gödel-complete would not imply that their known evident consequences are more extensive than those of a given Gödel-incomplete counterpart.

To the extent that this is so, Gödel completeness can only be seen as at best a secondary ideal of descriptive axiomatization. It is not generally required for descriptive adequacy – that is, for the completeness of an organizational scheme with respect to the data it is supposed to organize. Neither is it required for evidential completeness – that is, for full utilization of the evidence that is represented by the data. Neither, finally, is it required for analytical completeness.

[13] As a matter of intension, of course, descriptive and evidential completeness are distinct, even if, extensionally speaking, they should coincide.

What generally matters for descriptive axomatization is descriptive adequacy, and what matters for descriptive adequacy is completeness with respect to the data.

For foundational (as distinct from local) cases of descriptive axiomatization, analytical completeness and evidential completeness also matter. Neither of these, though, requires Gödel completeness (i.e., completeness with respect to the truth *per se*) either.[14] This being so, Gödel completeness seems to be of at most secondary importance as an ideal of descriptive axiomatization or as a condition for the realization of the more compelling such ideals.

6 Gödel completeness as a nondescriptive ideal

This notwithstanding, there may be other ideals of axiomatization (be they ideals of descriptive axiomatization or axiomatization of some other type) that Gödel completeness does serve. And it may be that an axiomatization that aims to be adequate descriptively also aims to satisfy certain of these ideals. I will now briefly consider this possibility, paying special attention to one traditionally influential possibility of this type – namely, the ideal of solving all mathematical problems. For convenience, I will refer to this as the ideal of *resolutive completeness*.

In the modern era, this was perhaps most closely associated with the name of Viète (cf. Viète 1983, §29), who presented it as the cardinal attraction of the use of what was then widely known as *the analytical art* (also as "algebra") to solve geometrical problems.

> [T]he analytical art . . . claims for itself the greatest problem of all, which is: TO SOLVE EVERY PROBLEM. (Viète 1983, p. 32)

More recently, and perhaps also more familiarly, resolutive completeness has been associated with the name of Hilbert, who presented it as follows:

> [N]owadays skepticism and faintheartedness are not infrequently expressed against science both in specialized publications and in public lectures . . . to the skeptics and the fainthearted I would say: *In mathematics there is no ignorabimus. On the contrary, we can always answer meaningful questions.* (Hilbert 1930b, p. 9, emphasis mine)[15]

---

[14] The differences between completeness with respect to the truth and completeness with respect to what is evident diminishes of course if the notion of truth involved is a constructive one. Here I am only considering Gödel completeness as an ideal with respect to generally realist, nonconstructivist conceptions of truth.

[15] Elsewhere (cf. Detlefsen 2005, §5.4), I have referred to the principle stated here as Hilbert's *axiom of solvability*.

It might be thought that acceptance of a principle of resolutive completeness would require acceptance of Gödel completeness as an ideal of mathematical theorizing. This is at least at first glance a reasonable view. Upon more careful examination, however, the two seem not to be so closely aligned.

There are different foci one might take in developing such an argument. Mine will be what I call the *solving-as-deciding thesis* (or SDT for short).

SDT: To solve a problem P?,[16] is to decide it convincingly – that is, it is to provide a convincing proof of P or a convincing proof of P's denial.[17]

6.1 The SDT: to solve P? is to decide it convincingly

The SDT was directly challenged by Hilbert as representing too narrow a conception of problem resolution. He maintained, in particular, that to prove that P cannot be decided by those methods under which its decision has traditionally and unsuccessfully been sought is also to solve P.

He mentioned a number of examples of this (e.g., the problem of squaring the circle, the problem of solving the general quintic by root extraction, etc., cf. Hilbert 1900a, p. 261) and characterized them as among the most outstanding (*hervorragende*) achievements in the history of mathematics. More pertinently for our purposes, he also described them as providing "fully satisfactory and rigorous (*strenge*) solutions" (Hilbert 1900a, p. 261) to the problems in question, "although in another sense than that originally intended" (Hilbert 1900a, p. 261).[18]

Another sense indeed. A sense too, though, which resonated with an important shift in the prevailing view of the nature of axioms and proofs that took place in the late nineteenth and early twentieth centuries. This is what I will call the *hypotheticist* shift.

---

[16] Here I am assuming that a problem can be expressed as a yes/no question P? and that it therefore represents the taking of an interrogative attitude (expressed by "?") towards the propositional content expressed by P.

[17] What constitutes a *convincing* proof is not something that need detain us here.

[18] This should not be taken to suggest that Hilbert was at that time skeptical of the existence of Gödel-complete theories. He was not. Nor was he generally skeptical in his later foundational writings. He made statements in fact which suggest that he believed both number theory and analysis to be Gödel-complete (cf. Hilbert 1930b, p. 6). In addition, he suggested that Gödel-completeness might make a suitable finitary substitute (i.e., a finitarily well-defined approximating notion) for the finitarily not well-definable notion of categoricity. It is less clear, I think, whether he himself saw such approximation as wanted for the purpose of accommodating a compelling ideal (for at least certain theories), or whether he saw it instead as a way legitimately to define and investigate approximations to ideals that had currency among other foundational thinkers of that time.

Axioms had traditionally been seen as self-evident truths that did not need, and, at least customarily, did not admit, of being proved. On the hypotheticist shift, this changed. Axioms were no longer considered to be evident, at least not by virtue of their office. They were instead regarded as hypotheses or assumptions which were to be logically explored.

Likewise, where proofs had traditionally been regarded as sequences of judgments by dint of whose observed logical connections one eventually obtained a warrant for the proof's conclusion, they were now seen as exposing logical relations between basic hypotheses and other propositions whose logical relationship to them was, for whatever reasons, a matter of interest to mathematical investigators.

Huntington put the salient points of the new view this way:

> The fact is that a mathematical demonstration, strictly speaking, is not concerned with the truth of the proposition at all; it is concerned merely with the logical relation that exists between the given proposition and certain other propositions called the axioms—in other words, all that a mathematical demonstration tells us is that if the axioms are true, then the theorem in question will also be true—provided, of course, that our deductive reasoning is sound. (Huntington 1911, §8)[19]

On such a view, what is essential to mathematical knowledge is knowledge of the logical connections that obtain – and, complementarily, knowledge of the logical connections that do not obtain – between propositions.

Some saw the shift to this view as having been rapid and dramatic, or at least relatively so. Some also saw it as a movement towards something like greater democracy in mathematical practice.

> The point of view of fifty years ago was very largely that the foundations of mathematics were axioms; and by axioms were meant self-evident truths, that is, ideas imposed upon our minds a priori, with which we must necessarily begin any rational development of the subject ... what is the new point of view? The self-evident truth is entirely banished. There is no such thing. What has taken the place of it? Simply a set of assumptions ... in the choice of which we have considerable freedom. The choice of a set of assumptions is very much like the election of men to office... they are elected for their fitness to serve, and their fitness is very largely determined by their simplicity, by the ease with which other propositions may be derived from them. (Young 1911, pp. 52–53)[20]

---

[19] See also Stewart (1814, §8 of the appendix), Peirce (1898, pp. 211–213) and Young (1911, pp. 51–53), to mention but a few of the many different sources and even general types of sources for such views.
[20] This is from a section of Young's essays entitled "The democracy of mathematics."

Whether the shift from the more traditional conception of axiom and proof was as Young describes it is not for my purposes important. What is important is the change in the basic conception of what constitutes a problem solution that such a shift signifies.

Traditionally, to solve a problem *P*? meant to decide *P* by evident means (i.e., to prove either *P* or its denial by evident means). With the hypotheticist shift it became something different, namely, to determine the logical relationships – including, possibly, the independence – between *P* and those not-essentially-to-be-regarded-as-evident hypotheses which, for whatever reasons, constitute the axioms with respect to which *P*? is being investigated.

Looked at another way, the hypotheticist shift changed the basic conception of what a problem is. Instead of a problem paradigmatically taking the form of a question concerning the truth of a proposition, it now came to have the form of a question concerning whether a given proposition can be deduced from certain other given propositions.

Hilbert described his *Grundlagen der Geometrie* as representing this type of undertaking.

> In this investigation the ground rule (*Grundsatz*) is to discuss every question that arises in such a way as to find out at the same time whether it can be answered in a specified way with some limited means. This ground rule seems to me to contain a general and natural prescription (*Vorschrift*)... if in the course of our mathematical reflections, we encounter a problem or imagine (*vermuten*) a theorem, our drive for knowledge (*Erkenntnistrieb*) is satisfied (*befriedigt*) only if either the complete solution (*völlige Lösung*) of the problem and the rigorous proof of the theorem are successfully demonstrated or the reason (*Grund*) for the impossibility of this success and hence the necessity of failure are clearly recognized. (Hilbert 1913, pp. 110–111)

The most meaningful form of completeness for any inquiry must surely be whether its execution satisfies that drive for knowledge for the sake of whose satisfaction it is undertaken. That knowledge, for axiomatic inquiry on the hypotheticist plan, is knowledge of whether and if so how a given proposition is decided by given specified means.

Judged from such a perspective, there is little reason to see resolutive completeness as requiring Gödel completeness, and there is good reason to say, as Hilbert said, that a proof of the independence of *P* from a given set of axioms can offer a "fully satisfactory and rigorous" (Hilbert 1900a, p. 261) solution of a problem *P*?.

It might be, though, that resolutory completeness would require, at the "limit" so to speak, complete identification of the logical connections between a proposition and a set of propositions. For cases of first-order consequence, such identification is not ruled out by Gödel's incompleteness theorem. It is in fact "ruled in" as a theoretical possibility by Gödel's completeness theorem for the first-order predicate calculus.

As is also well known, however, this reasoning does not extend to cases of second-order consequence. There, Gödel's *incompleteness* theorem (together with certain other facts) does matter. It prohibits even the theoretical possibility of complete identification of consequences.

As regards the relationship between Gödel completeness and resolutory completeness, then, the truth is not simple. On a hypotheticist conception of axiomatization, Gödel completeness is not generally necessary for resolutory completeness. It will be necessary only in those hypotheticist axiomatic settings where the identification of all instances of second-order consequence is the appropriate standard for resolutory completeness.

How extensive and significant the class of such settings might be, I am not sure. I have therefore made no claim concerning this. My claims are rather that the hypotheticist conception of axiomatization *is* significant, and that for axiomatizations conceived on the hypotheticist plan, Gödel completeness is not generally necessary for resolutory completeness. Cases where it might matter deserve closer consideration, but I do not have the space to go into this here.[21]

7 Conclusion

My aims have principally been (i) to call attention to certain differences between descriptive axiomatization and other types of axiomatization, (ii) to highlight the overall descriptive character of Whitehead's and Russell's axiomatization in the *Principia* and (iii) to offer a reassessment of the effects of Gödel incompleteness on the *Principia* (and similar) project(s) in light of these clarifications.

---

[21] I do not believe that Gödel completeness is generally a precondition for resolutory completeness, not even for axiomatizations conceived on the traditional plan. For it to be, some such condition as the following would be required.

> The ideal of solving every mathematical problem is essentially the ideal of developing, for each problem-area A of mathematics, a language L_A and a theory T_A such that T_A decides every sentence of L_A.

But such a condition does not seem to be justified. Generally speaking, there is no apparent reason why the solutions of all problems in a given area should be concentratable within the space of a single theory.

The argument underlying this reassessment pointed out that (a) for foundational descriptive axiomatization of the type represented by the *Principia*, the compelling completeness ideals are descriptive completeness, analytical completeness and regressive-evidential completeness, that (b) none of these completeness ideals seems to require Gödel completeness for its fulfillment and that (c) to extend a Gödel-incomplete theory T to a Gödel-complete theory T^+ would not seem to guarantee that T^+ be either an analytic extension[22] or an evidential extension of T.[23]

For descriptive axiomatization generally, then, and for the type of descriptive axiomatization represented by the *Principia* project in particular, Gödel completeness seems not to be a compelling ideal. Neither does it appear generally to be a condition whose satisfaction is necessary (or sufficient) for the fulfillment of such ideals.

It may of course be that, in addition to descriptive axiomatization, there are other comparably significant types of axiomatization for which Gödel completeness either is an ideal or is needed for the realization of such. This, though, is a topic for another occasion.

[22] Here by saying that T^+ is an *analytical extension* of T I would mean roughly that the axioms of T^+ extend the analysis of the data which terminates in T's axioms.

[23] In saying that T^+ is an *evidential extension* of T, I mean roughly that T^+ makes fuller regressive-evidential use of the data of T than T does. Here I am allowing that fuller use might consist in T^+'s being as well supported (in all its extended strength) by T's data as T is.

CHAPTER 5

Logical completeness, form, and content:
an archaeology

Curtis Franks

A thesis that is inconceivable now seemed obvious in the ninth century, and it somehow endured until the fourteenth century. Nominalism, which was formerly the novelty of a few, encompasses everyone today; its victory is so vast and fundamental that its name is unnecessary. No one says that he is a nominalist, because nobody is anything else.[1]

1 Les Mots et les Choses

The "signifying function" of words in the sixteenth century, according to Foucault, depends not upon our acquaintance with them or with their use, "but with the very language of things" (Foucault 1970, p. 59). Duret's (1613) etymology is essentialist: "Thus the stork, so greatly lauded for its charity towards its father and mother, is called in Hebrew *chasida*, which is to say, meek, charitable, endowed with pity ... " (Foucault 1970, p. 36). And conversely, the nature of a thing is prescribed by its formal features: its name, its shape. That ingesting a walnut will relieve a headache is written into the fruit, in its resemblance with the human brain (Foucault 1970, p. 26). "The names of things were lodged in the things they designated, just as strength is written in the body of the lion, regality in the eye of the eagle, just as the influence of the planets is marked upon the brows of men ... " (Foucault 1970, p. 36). The link does not require proof, is in fact closed off from critique. Aldrovandi, Grégoire, all the thinkers of the Renaissance, assumed such associations prior to their investigations. They were "meticulously contemplating a nature which was, from top to bottom, written," for they inhabited "an unbroken tissue of words and signs, of accounts and characters, of discourse and forms" (Foucault 1970, p. 40).

In the seventeenth century, we are told, things are different. Language is believed to be arbitrary, its relationship to the world contingent on the

---

[1] Jorge Luis Borges (1964), "From allegories to novels."

details of its fallible design and conventional use. "As a result," Foucault urges, "the entire *episteme* of Western culture found its fundamental arrangements modified. And, in particular, the empirical domain which sixteenth century man saw as a complex of kinships, resemblances, and affinities, and in which language and things were endlessly interwoven— this whole vast field was to take on a new configuration" (Foucault 1970, p. 54). When each sign had been assumed to be true to the world it resembled – when the world consisted entirely of signs – we faced the possibly endless task of reading all its naturally endowed features in order to know all that it expressed. But now each sign can be known in full and is, in fact, completely known, for it is an artifact precisely as the world is not. We can no longer wonder whether we have fully grasped the sign, for there is nothing in it that we did not put there. We are instead in doubt about what before could not meaningfully be questioned: whether our systems of signs adequately fit the world.

> The simultaneously endless and closed, full and tautological world of resemblance now finds itself . . . split down the middle: on the one side . . . the signs that have become tools of analysis, marks of identity and difference, . . .; and, on the other, the empirical and murmuring resemblance of things, that unreacting similitude that lies beneath thought . . . And between the two, the new forms of knowledge that occupy the area opened up by this new split. (Foucault 1970, p. 58)

When the question about the adequacy of language to the world recurs in the concrete terms of modern logic, the patterns of reconfiguration Foucault described are recognizable. The metamorphosis of logic from philosophical topic to mathematical subject was enabled by the identification of proofs with texts and the scientific study of tangible inscriptions. The written word can be surveyed, its properties observed. Questions of right reasoning, even of truth, can now be submitted to combinatorial tests. Logic is in its "period of transparency and neutrality" whereas before it had centered around the dependencies of thoughts and propositions on one another (Foucault 1970, p. 56). But to legitimize this scientific turn one must show that modern logical systems really do encode the rules of inference and criteria of truth that they leave behind in their "withdrawal from the midst of beings themselves" (Foucault 1970, p. 56).

Kurt Gödel answered this question for a fragment of logical language when he proved, in his thesis (1929), that "the restricted functional calculus" suffices to prove all universally valid first-order formulas. In 1930 Gödel said that the question he answered arises immediately for anyone who sets out to study this calculus (Gödel 1930c, reprinted in Gödel 1986, p. 103). Whether some such system could be shown to be "complete" in this sense had been explicitly posed over a decade before. Its analog was proved for a system of propositional logic around the same time: by Paul Bernays (1918) in his *Habilitationsschrift* and again by Emil Post (1921). In 1967 Gödel reflected on the fact that the technical achievements underlying his completeness theorem were present already in the work of Herbrand and Skolem, although neither writer managed to see that the result was essentially in his hands, and wrote to Hao Wang, "The blindness ... of logicians is indeed suprising" (Wang 1996, p. 240).

Gödel attributed the "blindness" of his predecessors to their lacking the appropriate attitude "toward metamathematics and toward nonfinitary reasoning" (Wang 1996, p. 240). Others have suggested that the principal obstacle in the way to the completeness theorem was a failure to appreciate the value of restricting one's attention to first-order quantification. "When Frege passes from first-order logic to a higher-order logic," van Heijenort (1967b, p. 24) writes, "there is hardly a ripple." Moore emphasizes instead the "failure" on the part of Gödel's contemporaries "to distinguish clearly between syntax and semantics" (Moore 1988, p. 126). While each of these accounts points to a substantial aspect of Gödel's characteristic approach to the study of logic, it is here suggested that "logical blindness" is reciprocating. Gödel never considered that others' logical vision might be, rather than defective, simply different – that their inability to see their way to the completeness theorem derived from their focus being held elsewhere. But thinking about logic in the terms that defined Gödel's contribution was not universal, perhaps not even common, in the early twentieth century. His early writing plays a major part in an implicit argument that the correspondence of proof and truth, of logical form and content, is a proper way of thinking about logical completeness. While the theorem contained in Gödel's thesis is a cornerstone of modern logic, its far more sweeping and significant impact is the fact that, through its position in a network of technical results and applications, the way of thinking underlying the result has come to seem definitive and necessary, to the extent that we have managed to forget that it has not always been with us.

The details of three separate notions of logical completeness, their histories, and the technical problems that they point to, should illustrate the contingency of the familiar conception of logical completeness as the correspondence of form and content. That conception, the one Gödel helped to drive into our basic understanding of logic, will be seen to have originated alongside a second conception based on a coordination of analytic and synthetic reasoning. Each of these modern conceptions can, moreover, be traced back to a common ancestor, a question about the relationship between the metaphysical dependence of truths on one another and the subjective lines of reasoning that lead from judgement to judgement. Far from diminishing the significance of Gödel's achievements, this new vision should showcase the profound conceptual clarification and reconfiguration of fundamental notions that operate between the lines of Gödel's thesis. Gödel's work is revolutionary not because of what questions he answered so much as because he managed, in answering them, to demonstrate the value of a particular way of asking them.

2 Bolzano's question

Bernard Bolzano engaged in the profound study of two distinct notions of logical consequence over several decades in the early nineteenth century. The work most remembered and highly regarded by modern logicians, because of its striking resemblance to twentieth century set-theoretical definitions of consequence, concerns the *Ableitbarkeit* ("derivability") relation. In his 1837 masterpiece, *Wissenschaftslehre*, Bolzano in fact defines a network of concepts – validity, compatibility, equivalence, and derivability – in terms of one another in a way very similar to contemporary presentations. Here is his definition of the last of these:

> Let us then first consider the case that there is a relation among the compatible propositions *A, B, C, D, . . . M, N, O, . . .* such that all the ideas that make a certain section of these propositions true, namely *A, B, C, D, . . .*, when substituted for *i, j, . . .* also have the property of making some other section of them, namely *M, N, O, . . .* true. The special relationship between propositions *A, B, C, D, . . .* on the one side and propositions *M, N, O, . . .* on the other which we conceive of in this way will already be very much worthy of attention because it puts us in the position, in so far as we once know it to be present, to be able to obtain immediately from the known truth of *A, B, C, D, . . .*the truth of *M, N, O, . . .* as well. Consequently I give the relationship which subsists between propositions *A, B, C, D, . . .* on the one hand and propositions *M, N, O, . . .* on the other the title, a relationship of *derivability [Ableitbarkeit]*. And I say that propositions *M, N, O, . . .* would be derivable from propositions *A, B, C, D, . . .* with respect to the variables *i, j, . . .*, if every set of ideas which makes *A, B, C, D, . . .* all true when substituted for *i, j, . . .* also makes *M, N, O, . . .* all true. (Bolzano 1837, §155)

Although this notion of derivability prefigures modern definitions of logical consequence in many ways, there are several evident disparities between

Bolzano's concept and our own. For one thing, Bolzano requires all the propositions involved in the *Ableitbarkeit* relationship to be "compatible" with one another. The result is analogous to a stipulation, absent from modern logical theory, that formulas be jointly-satisfiable in order to stand in a relationship of logical consequence with one another. One result of this unfamiliar requirement is that, for Bolzano, nothing at all is derivable from a self-contradictory proposition, whereas in modern logical theory all formulas are consequences of an unsatisfiable one. It is also noteworthy that Bolzano attends to propositions, not formulas, and to their reinterpretations over a fixed domain of ideas. This is a more conservative approach to modality than the modern one, wherein not only may the extensions of predicate symbols and constant symbols vary, but so too may the underlying set of objects. Furthermore, the individuation of "ideas" with respect to which one may vary one's interpretation is made imprecise by the focus on "propositions" and their constituents in place of the modern focus on formulas and the symbols they contain. Bolzano's stance on these matters, though it appears peculiar from a modern point of view, was not whimsical. He maintained his position consistently over many years. However, the modern notion is not conceptually distant from Bolzano's on these points and can meaningfully be seen as a refinement or adjustment of his definition.

Nevertheless, a strong contrast must be drawn between Bolzano's *Ableitbarkeit* relation and the modern notion of logical consequence, if not in terms of their technical details, in terms of the sort of relationship their authors took themselves to be defining. The logical consequences of a formula, on the modern view of things, are solely determined by the existence and details of certain set-theoretical structures, quite independently of our access to them or ability to draw inferences based on them. Bolzano's *Ableitbarkeit* relation, by contrast, is procedural. There is nothing "out there" over and above particular deductions that we might perform that determines any special relationship between the propositions that bear this relation to one another. In Bolzano's preferred terminology, the relationship corresponds to no "objective dependence" of propositions on one another (Bolzano 1810, II.§12). It is merely the case that *we* are able "to obtain immediately from the known truth of A, B, C, D, . . .the truth of M, N, O, . . .as well" and furthermore are able to do this in an a priori manner. When he wrote (1837, §155), "[fo]r the sake of variety, I shall also sometimes say that propositions M, N, O, . . . *follow from* or can be *inferred* or *concluded* from the set of propositions A, B, C, D, . . .," he did not explain the phenomenon of inferring correctly in terms of a metaphysical relationship between propositions that our inferences might track. He treated right reasoning as primitive, the variation of ideas in propositions as part of the inferential process.

Of course, Bolzano was not claiming that propositions only stand in the *Ableitbarkeit* relation with one another after someone has in fact carried out a logical deduction. It is an objective and eternal fact, for Bolzano, whether or not such a relationship attains. So what is his point in saying that the relationship is only subjective, that a truth obtained in this way is a "mere conclusion [*bloßer Schlußsatz*]" and not a "genuine consequence [*eigentliche Folge*]" (Bolzano 1837, §200)?

According to Bolzano, the fact of propositions *A, B, C, D, ...* standing in the *Ableitbarkeit* relation to propositions *M, N, O, ...* comes to no more than our ability rightly to infer the latter from the former. The correctness and incorrectness of our inferences is objective. It is typically erroneous, however, to call the latter the consequence of the former, because one proposition being the consequence of another is itself a feature of reality independent of what knowledge we may have of this fact and the use, in reasoning, we might make of it. Indeed, throughout his logical investigations, Bolzano's considerably more sustained focus was devoted, not to the *Ableitbarkeit* relation, but to the theory of this objectively significant consequence relation, a theory he called "*Grundlehre.*"

Bolzano's 1810 *Beyträge* is the definitive exposition of this theory of ground and consequence. In §2 of part II of that booklet, Bolzano wrote:

> in the realm of truth, i.e. in the sum total of all true judgements, a certain *objective connection* prevails which is independent of our actual and *subjective recognition* of it. As a consequence of this some of these judgements are the grounds of others and the latter are the consequences of the former. To represent this objective connection of judgements, i.e. to choose a set of judgements and arrange them one after another so that a consequence is represented as such and conversely, seems to me to be the proper *purpose* to pursue in a scientific exposition. Instead of this, the purpose of a scientific exposition is *usually* imagined to be the greatest possible *certainty* and *strength of conviction.*

This consequence relation, which Bolzano called *Abfolge*, is at the center of a robust philosophical account of mathematical truth. Its influences are legion. A proof, according to Bolzano, must track the objective *Abfolgen* between propositions. Individual mathematical truths therefore have at most one proof (Bolzano 1810, §5). Moreover, the division of mathematical truths into those that have proofs and those basic truths for which no proof can be given is not a matter of convention but is objectively determined and there for us to discover (Bolzano 1810, §13). The distinguishing features of an axiom are, accordingly, not its self-evidence, but its onto-logical role as ground for other truths and the absence of any proposition serving in the capacity of its ground (Bolzano 1810, §14). Conversely, and most importantly for Bolzano, the self-evidence of a mathematical fact is no reason not to seek a proof for it, for a proof will uncover its grounds, which are typically unrelated to the (good) reasons we might have for accepting the fact as true (Bolzano 1810, §7).

One might reasonably wonder why this hypothesized network of objective relationships should more properly be the focal point of "scientific expos-ition" than the simple discovery of mathematical facts. Bolzano provided several justifications for the shift in perspective that will serve to illustrate further his concept of consequence.

Primarily, and most often, Bolzano points to an inherent value in coming to understand the structure of the hierarchy of facts. This hierarchy is a feature of the world forever off limits to researchers who "stop short" at certainty. Behind this incentive is the idea that proofs, of the special sort that Bolzano seeks, are explanatory: a fact's grounds are the reason why that fact is true. In some sense they constitute their consequences, and therefore being more than "*Gewissmachungen*" that assure us of a truth, proper proofs are "*Begründungen*, i.e., presentations of the objective reason for the truth concerned" (Bolzano 1817b, Preface, §I). Science should not simply record but should also explain facts.

There is also an aesthetic value to Bolzano's proofs. Through them, one is able to see one's way to a mathematical truth without recourse to ideas and terms that are "off topic." "[I]f there appear in a proof *intermediate concepts* which are, for example, *narrower* than the subject, then the proof is obviously defective; it is what is usually otherwise called a μετάβασις εἰς 'άλλο γένος"[2] (1810, II. §29). Thus, although Bolzano did not actually define the *Abfolge* relation or specify, in any but a few select cases, what the unprovable basic truths are, he disclosed a highly nontrivial fact about the *Grundlehre*: every nonbasic fact is grounded in other facts about one and the same concepts that the consequent, nonbasic fact is about.

Bolzano further hinted that the conceptual purity of his proofs affords a scientific advantage, in that it will facilitate the discovery of new truths. In his 1804 *Betrachtungen*, after claiming that "one must regard the endeavor of unfolding all truths of mathematics down to their ultimate

---

[2] "Crossing to another kind."

grounds ... as an endeavor which will not only promote the *thoroughness* of education but also make it *easier*," Bolzano wrote:

> Furthermore, if it is true that if the first ideas are clearly and correctly grasped then much more can be deduced from them than if they remain confused, then this endeavor can be credited with a *third* possible use—the *extending* of the science. (Bolzano 1804, paragraph 3)

In the continuation of this passage and again in the *Beyträge* he recounted episodes in the history of mathematics when attempts to prove facts about which there was already no doubt led to the discovery of new truths.

Bolzano's first justification of the centrality of the *Abfolge* relation strikes modern readers as arcane. As Rolf George has remarked, "It seems absurd to try to decide whether two lines intersect because they share a common point, or have a point in common because they intersect," and yet Bolzano insisted that "nobody who has a clear concept of ground and consequence will deny that the first proposition is not objectively grounded in the second, but the second in the first" (George 1971, p. xxxvii). Meanwhile, the second justification seems ideological and the third speculative. Are impure proofs really defective in some way? Are pure proofs more explanatory? Might the seemingly disorganized methods of expert mathematicians not actually have some scientific advantage over Bolzano's contrived efforts to track the objective dependencies of truths? Bolzano did not address these individual points in a way that would persuade many skeptics. However, he emphasized a fourth advantage to his standard of proof with greater scientific promise: if we are not careful in mathematical reasoning to uncover the objective grounds for our claims, then the impurity of our demonstrations will lead to disorder, threatening circularity in our reasoning. "For example," Bolzano wrote,

> in [Lagrange's] *Théorie des fonctions analytiques, No. 14*, the important claim that the function $f(x + i)$, with i continuously variable, can in general be expressed
>
> $$f(x + i) = (x) + ip + i^2 q + i^3 r + \cdots$$
>
> is derived from a *geometrical* consideration: namely from the fact that a continuously curved line which cuts the *x*-axis has no smallest ordinate. Here one is in a real *circulus vitiosus*, because only on the assumption of the purely arithmetical assertion about to be proved can it be shown that every equation of the form $y = fx$ gives a continuously curved line. (Bolzano 1810, §29)

The same point appears to be the impetus of his celebrated proof (Bolzano 1817b) of the intermediate value theorem. He began that paper by examining the geometrical demonstrations of this theorem by Euler, Laplace, and others, and observed that "such a geometrical proof is, in this as in most cases, really circular. For while the geometrical truth to which we refer is . . ., extremely *evident*, and therefore needs no *proof* in the sense of *confirmation*, it none the less needs *justification*." This alone does not substantiate his charge of circularity. But Bolzano proceeded to ask us to consider "the objective reason" why that geometrical truth attains and remarked:

> Everyone will, no doubt, see very soon that this reason lies in nothing other than that general truth, as a result of which every continuous function of x which is positive for one value of x, and negative for another, must be zero for some intermediate value of x. And this is just the truth which is to be proved. (Bolzano 1817b, Preface, §I)

Bolzano was thus not simply after a glimpse at the alleged causal links among truths. There is a clear scientific reason why the theorems of mathematics should be justified through an analysis that discloses their objective grounds. Because such an analysis will track the objective dependencies of truths on one another, there will be no risk that the evidence such analysis uncovers for a truth will itself depend, in a circular fashion, on that truth. When he examined Euclid's *Elements* he observed the "dissimilar objects" dealt with therein:

> Firstly triangles, that are already accompanied by circles which intersect in certain points, then *angles*, adjacent and vertically opposite angles, then the *equality* of triangles, and only much later their *similarity*, which however, is derived by an atrocious detour [*ungeheuern Umweg*], from the consideration of parallel lines, and even of the *area* of triangles, etc.! (Bolzano 1810, Preface)

The Euclidean method is thus unsatisfactory by Bolzano's lights, but because it draws its sting from the threat of circularity, this dissatisfaction is no mere matter of preference. For Bolzano it is a call for bold revision in mathematics: "But if one considers," he continued, how well-composed the *Elements* is, "and if one reflects how every successive proposition, with the proof with which Euclid understands it, necessarily requires that which precedes it, then one must surely come to the conclusion that the reason for the disorder must be fundamental: the entire method of proof which Euclid uses must be incorrect."

In developing the *Grundlehre*, Bolzano advanced logical theory in ways comparable in scope to his work on *Ableitbarkeit* but oriented in a different direction. The initial insight seems to have occurred to him rather early. In his youthful 1804 pamphlet he wrote, "I must point out that I believed

I could not be satisfied with a completely strict proof *if it were not even derived from concepts* which the thesis proved contained, but rather made use of some fortuitous alien, *intermediate concept [Mittelbegriff]*, which is always an erroneous μετάβασις εἰς 'ἄλλο γένος³" (Bolzano 1804, Preface, paragraph 4). That proofs should be free from such intermediate concepts and the concomitant "atrocious detours" in reasoning (attributed to Euclidean methods) was inspired by the desire to capture the objective ground and consequence relations in the world, to produce proofs that were topically pure and therefore free from circularity. However, the significance of the notion of analyticity that Bolzano developed is not tied down to those ambitions.

In §17 of part II of the *Beyträge* (1810), Bolzano had distinguished analytic and synthetic truths according to the Kantian criterion of conceptual containment (the predicate of an analytic, and not a synthetic, truth contains its subject). In §31 he extended this to a distinction between analytic and synthetic proofs. He did not explicitly define the notions, but they can be reconstructed from context: a proof is analytic if its derived formula contains, in its compound concepts, all the simple concepts that appear elsewhere in the proof. Remarkably, Bolzano suggested that "the whole difference between these two kinds of proof [analytic and synthetic] is based simply on the *order and sequence* of the propositions in the exposition." Thus Bolzano rediscovered the formidable ontological burden that he placed on proofs reflected in a rather mundane feature of the written appearance of those proofs.

This observation is supported by the rudiments of a theory of proof transformation, outlined in §20. Because every compound proposition is built out of a subject and predicate which depend on the individual concepts of which it is composed, the proposition itself, if true, "is actually also a derivable, i.e. provable proposition." Moreover, its single proof begins with only simple propositions about the simple concepts contained in the compound, proved proposition. One of Bolzano's great discoveries is that the rules of inference in a proper analytic proof that lead from these simple propositions to the proved proposition are other than the patterns of syllogistic reasoning to which logicians in his day devoted so much attention. To make this point, he embarked on a fine classification of inference rules:

> it should now be discussed at length *how many simple and essentially different kinds of inference there are*, i.e. how many ways there are that a truth can

³ "Crossing to another kind."

depend on other truths. It is not without hesitation that I proceed to put forward my opinion, which is so very different from the usual one. Firstly, concerning the *syllogism*, I believe there is only a single, simple form of this, namely *Barbara* or γραμματα in the *first* figure. (Bolzano 1810, §12)

Bolzano suggested a modification of *Barbara*'s classical presentation, based on the natural way of reasoning and continued, "However, these are small matters.—Every other figure and form of the syllogism seems to me to be either not essentially different from *Barbara* or not completely simple." There is no explicit presentation in the *Beyträge* of criteria for simplicity and identity of inference rules, but clearly suggested in these remarks is the idea that *Barbara* in the first figure is all one would have to use in order to derive any proposition derivable from the full gamut of classical syllogistic forms. These ideas alone are substantial in the history of logic, but what Bolzano proceeded to describe is even more so. "But on the other hand," he wrote, "I believe that there are some *simple kinds of inference* apart from the syllogism." Among his examples is the inference from "*A is* (or *contains*) *B*" and "*A is* (or *contains*) *C*" to "*A is* (or *contains*)[*B* et *C*]." "[I]t is also obvious," Bolzano claimed "that according to the necessary laws of our thinking the first two propositions can be considered as *ground* for the third, and not conversely" (Bolzano 1810, §12).

After illustrating a couple of other such rules, which similarly establish compound clauses within the sub-sentential structure of propositions, Bolzano noted a crucial difference between his new "analytical" rules and the syllogism, clearly based on his rich notion of *Abfolge*. The syllogism rule is not reversable – its premises in no way follow from its conclusion – but the analytical rules each are. For this reason, the reverse of each analytical inference "could seem like an example of another kind of inference . . ."

> But I do not believe that this is a [proper] *inference* . . . I can perhaps *recognize* subjectively from the truth of the *first* of these three propositions the truth of the two others, but I cannot view the first *objectively* as the *ground* of the others. (Bolzano 1810, §12)

Thus propositions with compound concepts can be proved in a way that charts the *Abfolge* hierarchy, i.e., purely analytically, from propositions containing only simple concepts. "On the other hand," Bolzano wrote in a long note to §20, "how propositions with simple concepts could be proved other than through a syllogism, I really do not know."

In §27, drawing from the observed features of the analytical inference rules, Bolzano argued for the following claim: "If several propositions appearing in a scientific system have the same subject, then the proposition with the more compound predicate must follow that with the simpler predicate and not conversely." His "proof" is just one line, after which he wrote, "This truth has forced itself particularly clearly upon those who have thought about the nature of a scientific exposition." He continued:

> Moreover, it is obvious here that we cannot extend our assertion *further*, and instead of the expression, "*the proposition with the more compound predicate*," put the more general one, "*the proposition with the narrower predicate*." For whenever we make an inference by a syllogism one of the premises (namely the so-called *major*), with just the same subject as the conclusion, has the predicate (namely the *terminus medius*) which is narrower than that of the conclusion: *S contains M, M contains P*, therefore *S contains P*, where the concept *M* must obviously be narrower than *P* because otherwise the proposition, *M contains P*, could not be true, and yet the judgement, *S contains M*, must be considered as one preceding the judgement, *S contains P*. (Bolzano 1810, §27)

In other words, Bolzano recognized that his proofs, because their propositions are ordered so as to track the objective *Abfolgen* in the world, would have a form of what modern logicians call a subformula property were it not for the ubiquity of the syllogism rule. Even with this rule, though, every proof has a related property. Given the normalizing techniques discussed in §20, typical proofs may generally be written so that they begin with several syllogisms devoted to establishing the needed simple truths from which to infer, purely from analytic rules, their more compound consequence. In §30 Bolzano touched on this again. There he seems to be saying that even within the syllogistic part of proofs, there is a determined single way to proceed.

Bolzano (1817a, preface) described a "purely analytic procedure" differently, as one in which a derivation is performed "just through certain changes and combinations which are expressed *by a rule completely independent* of the nature of the designated quantities." This description points to the features of analyticity emphasized by the eighteenth century algebraists, who sought to extend algebraic techniques to mechanics, geometry, and other disciplines. Laplace (1813), for example, in chapter 5 of book V of his *Exposition du Système du Monde* had written:

> The algebraic analysis soon makes us forget the main object [of our researches] by focusing our attention on abstract combinations and it is only at the end that we return to the original objective. But in abandoning oneself to the operations of analysis, one is led by the generality of this method and the inestimable advantage of transforming the reasoning by mechanical procedures to results inaccessible to geometry. (Kline 1972, p. 615)

Similarly Lagrange (1788, preface) declared, "The methods which I expound [in *Mécanique analytique*] demand neither constructions nor geometrical or mechanical reasonings, but solely algebraic operations subjected to a uniform and regular procedure." Nowadays one reflexively associates these "mechanical procedures" with derivability and conceives of logical consequence as residing in the "object of our researches," on the semantic half of this divide. One wonders whether one's formulas are adequate to their intended interpretations, whether these mechanical procedures in fact trace the interrelationships among the objects of our researches, these latter being "the original objective." Bolzano could only have it the other way around. His *Grundlehre* revealed that the analytic calculus traces facts back to their ultimate, constitutive grounds, and he shared Laplace's suspicion that these same dependencies might be inacessible by geometrical or other traditional mathematical inferences. Should one side of this divide prove inadequate, it could only be the latter – this being rightly designated as mere derivability – because "by abandoning oneself to the operations of analysis" one accesses the objective dependencies among truths.

Bolzano's concepts of *Ableitbarkeit* and *Abfolge* each deserve detailed reconstruction and appreciation. Their influence on our current understanding of logic is significant, yet unlikely fully known. Doubtless, too, in serving aims that no longer drive modern investigations – precisely, that is, where they have failed to be influential – they are windows into possible lines of thought that history, for one reason or another, simply did not preserve. Perhaps their greatest legacy lies in neither of them individually, though, but in Bolzano's desire to establish a correspondence between them.

Whereas the *Abfolge* relation holds only between truths, false propositions may stand in the relationship of *Ableitbarkeit* with one another so long as (1) under some substitution of ideas they all are true and (2) under all substitutions that make some one part of them true, so too is the second part. For this simple reason, one cannot conclude from the derivability of some proposition that one has uncovered the grounds, in the premises of this derivation, of a proposition. More crucially, the same conclusion cannot be drawn even when all the propositions in the derivation are true. This is evident from the fact that derivability is obviously reflexive and often symmetric, whereas according to Bolzano no truth is its own ground (1837, §204), and no two truths could mutually ground one another (§211). The *Ableitbarkeit* relation is not, in Bolzano's idiom, "subordinate" to the *Abfolge* relation.

Observing this gap between his two notions of consequence, Bolzano wrote, in §162 of his *Wissenschaftslehre*:

Not every relationship of *derivability*, then, is so constituted that it also expresses a relationship of ground to consequent holding between its propositions when they are all true. But the relationship of derivability that does posses this characteristic will doubtless be sufficiently worthy of attention to deserve a designation of its own. I will therefore call it a *formal ground-consequence* relation [*formalen Abfolge*] . . . (Bolzano 1837, §162)

Given the nature of these relations, the converse question seems more relevant: is every *Abfolge* relation representable with a derivation? In the terms Bolzano coined, is every *Abfolge* relation in fact a *formalen Abfolge* relation? If not, then the very idea of placing this conception of logical consequence at the center of all scientific exposition is puzzling. Whatever the merits of knowing the objective grounds of a mathematical fact, no science can be devoted to this task without some sort of method for the discovery of such grounds. Yet Bolzano managed only to indicate a handful of unrelated instances of this fundamental relation, appealing haphazardly to theological, moral, and intuitive ideas to do so. On the other hand, if from its grounds a truth can always, in principle, be derived, then the process-centered *Ableitbarkeit* relation is seen to be adequate to trace the objective dependencies of truths on one another. This alone would not establish that the theory of derivability would allow us to discover the grounds of every proposition. The problem, again, is that far too many things are similarly derivable. But it would guarantee that the objective consequence relation does not lie beyond the reach of human inferential practices. All such consequences would be captured in a derivation, and from further considerations about the features of a derivation – whether, for example, it unfolds into an analytic proof – one would be able to determine whether it has turned up the ultimate reasons for the proposition's truth.

Bolzano devoted §200 to this question. The section is entitled, "Is the relation of ground and consequence subordinate to that of derivability?" Here is how he explained the point:

> If truths are supposed to be related to each other as ground and consequence, they must always, one might believe, be derivable from one another as well. The relation of ground and consequence would then be such as to be considered a particular species of the relation of derivability; the first concept would be subordinate to the second. (Bolzano 1837, §200)

Tempting as this belief might be initially, after a little reflection it is untenable. Why should the ultimate reasons for the myriad truths of mathematics always be related to them so that from consideration of the variation of ideas in each, one can reliably infer from them their objective dependencies? Why should the formal features of propositions give us any access to the shimmering reality beyond? Few contemporary writers share Bolzano's confidence in our intuitions about the realm of objective dependencies, but even Bolzano recognized the unconvincingness of speculation on this issue. "Probable as this seems to me," he concluded, "I know no proof that would justify me in looking upon it as settled." One can diagnose the difficulty more definitely, as not simply the absence of a proof but indeed the lack of any clear idea of what a proof might look like. Bolzano's two theories of logical consequence are themselves not precise enough for their correspondence with one another to be subject to proof. All the same, the question is at the center of Bolzano's thought. The procedural *Ableitbarkeit* relation provides a calculus of inference. The ontological *Abfolge* relation is a feature of the world absolutely independent of our ability to reason about it. By establishing that these notions correspond, we would ensure that the logical structure of the world is accessible, that some line of thought could trace the dependencies of truths, that the reasons behind the complex facts of reality are discoverable and comprehensible.

3   Gentzen's answer

The single point at which all of Bolzano's logical investigations are focused is, from the modern point of view, deeply suspect. Logic is blind to considerations of truth, to say nothing of ultimate explanations for why some statements are in fact true. Indeed, Bolzano's own development of *Ableitbarkeit* was a turn away from factual truth, towards distinguishing those statements that could be true from those that could not, towards identifying statements that rise and fall together no matter what the world is like. "Mathematics," he wrote, "concerns itself with the question, *how must things be made in order that they should be possible?*" unlike metaphysics which "raises the question, *which things are real?*"(Bolzano 1810, I. §9). But in the end, it was the objective grounding of truths that drove him, and if the theory of *Ableitbarkeit* cannot be shown to trace the world's *Abfolgen*, it loses much of its scientific interest. Modern logicians, by contrast, have no expectation that their craft will uncover ultimate grounds. Many do not even believe in such things. What remains of Bolzano's intricate scheme for writers who do not share his metaphysical aspirations?

In the century after Bolzano's work, logicians managed to uproot nearly the whole apparatus that he developed from its ontological setting. Astonishingly, not only did they preserve many of the details of his theory in the process, but the metaphysical character of Bolzano's thought proved in many ways to be the principal obstacle to its development.

One peculiarity of the dilemma at the center of Bolzano's logic is that Bolzano's own development of the *Grundlehre* revealed a strikingly manageable system of analysis, whereas the reportedly inference-based theory of *Ableitbarkeit* is quite difficult to implement. The proofs in the *Grundlehre* are chains of applications of a single syllogism rule together with a collection of analytical rules for "combining concepts." Given a true proposition, the task of building its proof-tree is generally not daunting. One merely identifies the complex concepts that it contains and looks up the rules needed to establish such a concept in the subject or predicate position. The immediate grounds of this proposition can then be constructed from the template that the relevant rule provides. Those grounds will themselves be truths, so their immediate grounds can be discovered in the same fashion. The only reason this procedure is not a fully adequate means for discovering a proposition's grounds is that often no straightforwardly analytic rule is applicable. Occasionally, as Bolzano noted, all the concepts in a truth are simple, and the way from it up to its ultimate grounds can only be charted by a syllogism. Unlike the rules that Bolzano introduced, the syllogism does not provide any hints as to the premises needed for a given consequence. As Bolzano explained, the predicate of the major premise and the subject of the minor premise, the so-called *terminus medius*, is not contained in the consequence and can only be guessed at. This is a serious obstacle, of course, to a fully mechanizable system of proof. But even partial mechanizability, of the sort provided by the analytical rules of inference, puts the *Abfolge* relation more within our reach than the relation of *Ableitbarkeit*. To determine "derivability," one must be ingenious at every turn. Which ideas ought one to vary? Which new ideas ought one to replace them with? And how, more dauntingly, can one tell that in a particular case all such replacements will yield the same result?

It is thus odd to consider the *Ableitbarkeit* relation as anthropomorphically grounded and the *Abfolge* relation as somehow beyond us. This categorization only makes sense given Bolzano's metaphysical distinction between *Gewissmachungen* and *Begründungen*, for the systematic study of each relation reveals that the former cannot feasibly be discovered, whereas discovering the latter is often largely a matter of routine calculation. Unhindered by Bolzano's emphasis that the *Abfolge* relation obtains only among truths, one might sensibly refer to this as a derivability relation and Bolzano's *Ableitbarkeit* as a definition of logical consequence. Such, at least, was the way things seemed to the twentieth century logician Gerhard Gentzen.

Gentzen is remembered primarily for two deep contributions to logic: the design of logical proof calculi that make perspicuous the flow of reasoning in a proof and the use of transfinite induction to calibrate the consistency strength of formal mathematical theories. It is fairly well known how these achievements relate to one another. Gentzen's logical calculi are susceptible, because of their division of inference rules into structural and analytic sorts, to tightly controlled proof-transformations. One can thus reason about the provability of certain formulas by considering how to transform alleged proofs of such formulas into certain canonical forms. Transfinite induction is used to track the transformation, and the perspicuous features of the canonical forms allow one to rule out the possibility of such a proof being written down. Thus, for example, if one thinks of consistency as the unprovability of a contradiction, one can determine that a mathematical theory is consistent by reflecting on the fact that any proof in that theory of a contradiction could be rewritten in a special, but obviously impossible, form.

Underlying Gentzen's logical achievements, however, is a more fundamental contribution. He introduced a notion of logical completeness that simultaneously made possible these scientific results and solidified a conception of logic completely dissociated from Bolzano's ontological scheme. Gentzen's idea is observable in his dramatic recasting of Bolzano's question.

Gentzen's conception of logical completeness originates in his first paper from 1932. The "formal definition of provability" in that paper consists of "sentences" of the form $M \to v$, where v is an "element" and M is a "complex" (a nonempty set of finitely many elements). Letting concatenation of letters represent their set-theoretical union, sentences can also be written with the elements of a complex displayed: $u_1 u_2 \ldots u_n \to v$. Gentzen referred to the complex at the left of a sentence's arrow symbol as its antecedent and to the lone element on the right of the arrow symbol as the succedent.

Prior to Gentzen, much attention was devoted to the distinction between categorematic and syncategorematic terms – the parts of language that signify on their own and those "logical particles" that serve merely to bind significant bits of language together so that they signify, not singly, but as a composite. A convincing definition of this distinction proved elusive but, it was thought, a fully articulated notion of logical consequence depended on such a definition. Even writers who shifted the locus of logical relationships from sentences to propositions sought some such distinction. The logical consequences of a proposition are those propositions that follow from it by dint of their structure, this structure being determined by their syncategorematic parts. Consider Bolzano's definition of *Ableitbarkeit*: "the relation among the compatible propositions $A, B, C, D, \ldots M, N, O, \ldots$ such that all the ideas that make a certain section of these propositions true, namely $A, B, C, D, \ldots$, when substituted for $i, j, \ldots$ also have the property of making some other section of them, namely $M, N, O, \ldots$ true." To determine whether propositions stand in this relation with one another, one has to know how finely one can carve them up into "ideas" and also how the parts of the propositions that are not "ideas" do their binding. As one burrows ever more deeply into a proposition, one finds sentential connectives, modal qualifiers, sub-sentential particles (quantifiers, etc.), and with each discovery the specification of a proposition's truth conditions, in terms of the contribution that these particles make to its meaning, is increasingly complex (and subject to debate). "To be sure," Bolzano wrote, "this distinction has its ambiguity, because the domain of concepts belonging to logic is not so sharply demarcated that no dispute could ever arise over it" (1837, §148).

Gentzen's response to this conundrum was to abandon the search for any final word as to the "logical" parts of a sentence and to develop a fully general account of logical consequence independent of such considerations:

> We say that a complex of elements *satisfies* a given sentence if it either does not contain all antecedent elements of the sentence, or alternatively, contains all of them and also the succedent of that sentence... We now look at the complex K of all (finitely many) elements of $p_1, \ldots, p_\nu$ and q and call q a *consequence* of $p_1, \ldots, p_\nu$ if (and only if) every subcomplex of K which satisfies the sentences $p_1, \ldots, p_\nu$ also satisfies q. (Gentzen 1932, p. 33)

Gentzen did not propose that the very simple structure of his sentences captured everything of logical importance about how propositions are related to one another. On the contrary, he wanted his sentences to display minimal logical structure – just enough to unpack meaningfully the basic notion of logical consequence. That notion, he claimed, does not depend on what the elements of a sentence are, and therefore ought to be studied in a setting that pried no more deeply into the structure of sentences than necessary. The result is at once free of controversy and applicable to a wide variety of topics. Gentzen suggested a few: interpret the elements as events and read $u_1 u_2 \ldots u_n \to v$ as "the happening of events $u_1 u_2 \ldots u_n$ causes the happening of v"; interpret the arrow as a containment relation and read the same sentence as "any collection that contains $u_1 u_2 \ldots u_n$ also contains v"; read the same sentence as "an object with the properties $u_1 u_2 \ldots u_n$ also has the property v." Among these he also mentions the more expected reading, "if the propositions $u_1 u_2 \ldots u_n$ are true, then the proposition v is also true."

Gentzen remarked, "Our considerations do not depend on any particular kind of informal interpretation of the 'sentences', since we are concerned only with their formal structure," and evidently the modicum of formal structure displayed in his "sentences" suffices for an adequate definition of our shared concept of logical consequence (Gentzen 1932).

It is fair to ask why Gentzen did not treat sentences themselves as elemental and say simply that one sentence is the consequence of some others if in any situation in which the latter are all true, so too is the former. The problem with this approach is that the informal notion that one arrives at cannot be "formalized." Gentzen viewed the above definition, not as a semantic analysis of the notion of logical consequence, but as a clarification of that informal notion preliminary to its formalization. There is, for example, no distinction in kind between the complexes that might satisfy a sentence and the parts of that sentence. In place of a theory about these complexes and their elements, Gentzen designed a logical proof system that formally captures the intuitive notion. He specified two inference rules for his system, which he called "THINNING" and "CUT":

$$\frac{L \to v}{ML \to v} \, thinning \qquad\qquad \frac{L \to u \qquad Mu \to v}{LM \to v} \, cut.$$

Then he defined a "proof" of a sentence q from the sentences $p_1, \ldots, p_v$ to be "an ordered succession of inferences (i.e., THINNINGs and CUTs) arranged in such a way that the conclusion of the last inference is q and that its premises are either premises of the p's or tautologies" (Gentzen 1932, reprinted in Szabo 1969, p. 31).

In section 4, Gentzen wrote:

> Our formal definition of provability, and, more generally, our choice of the forms of inference will seem appropriate only if it is certain that a sentence q is "provable" from the sentences $p_1, \ldots, p_v$ if and only if it represents informally a consequence of the p's. (Gentzen 1932, reprinted in Szabo 1969, p. 33)

In the first two theorems of the paper, Gentzen proved just this. In fact he showed that proofs of a specific "normal form" (chains of applications of CUT followed by a single, terminal application of THINNING) suffice to exhibit all the consequences among sentences.

This proof system thus formally realizes the concept of logical consequence in a way that is abstracted from considerations about the specific features of propositions by virtue of which they follow from one another. Because of the normal form component of Gentzen's proof, one can in fact say more.

The logical consequences of a set of sentences are each derivable from them with the single rule CUT, perhaps followed by a weakening of the derived assertion with the rule THINNING. In other words, Gentzen showed that CUT is the formal inference rule underlying the intuitive consequence relation.

In his 1932 paper, Gentzen applied this proof system to questions about the independent axiomatizability of theories. The deeper significance of Gentzen's "formal definition of derivability," however, is that it provides a setting for a precise formulation of a notion of logical completeness. This is the use that Gentzen made of his system in his *Untersuchungen* (1934–1935). The idea is simple. First replace the elements of the proof system with formulas from a specific branch of logic. In the *Untersuchungen* Gentzen studied the first-order predicate calculus. Then add to the "structural rules" CUT and THINNING, new rules associated with the logical particles of these formulas. These "logical rules" formalize the inferences that lead to and from propositions containing the associated logical particles. The resulting system is called a "sequent calculus." Gentzen illustrated how he arrived at his logical rules from an empirical study of mathematical proofs.[4] But the question immediately arises, whether or not rules arrived at in this way fully capture the meanings of the logical particles they are associated with. Can one extract with these rules every logical consequence of some premises? If so, then it is reasonable to say that the rules associated with the logical connectives are complete in the sense that they fully capture those connectives' meanings – but how can one show that this is the case? The sequent calculus allows this question to be posed precisely: Can everything derivable in fact be derived without CUT, or is the CUT rule an essential ingredient in some logical derivations?

In §3 of the *Untersuchungen* Gentzen answered this completeness question for both classical (LK) and intuitionistic (LI) sequent calculi formulations of predicate logic with his famous "*Hauptsatz*": "Every LI- or LK-derivation can be transformed into another LI- or LK-derivation with the same endsequent, in which no CUTs occur." It is noteworthy that the LI and LK results are essentially the same – these calculi differ only in the structure of their sequents – whereas other conceptions of completeness, wherein syntactic calculi are coordinated with semantic theories, require essentially different

---

[4] Gentzen first codified empirically observed inference rules in his natural deduction calculus. The sequent calculi rules are the result of rewriting the natural deduction rules so that the analytic rules associated with logical particles become disentangled from the synthetic operation DETACHMENT or, in the final analysis, CUT.

proofs for classical and intuitionistic logic. It is furthermore noteworthy that several features of Gentzen's logical writing indicate that he viewed the *Hauptsatz* as a completeness result. That evidence is presented in Franks (2010). One detail relevant to the current discussion is the fact that although Gentzen worked after, and in full awareness of, Gödel's results from 1929–1931, he seemed not to appreciate or acknowledge Gödel's own completeness theorem.

Gentzen's formulation of logical completeness has an additional theoretical feature that results from the nature of the logical rules in his system. Those rules are "analytic" in the same sense that the rules of Bolzano's *Grundlehre* are. Their premises contain no ideas that their conclusion does not contain. In Gentzen's modern terminology, the formulas in a premise of such a rule are all subformulas of the formulas in its conclusion. Every formula in a derivation without the CUT rule therefore occurs as a subformula in the endsequent. Gentzen put the point like this:

> these properties of derivations without cuts may be expressed as follows: [Their formulas][5] become longer as we descend lower in the derivation, never shorter. The final result is, as it were, gradually built up from its constituent elements. The proof represented by the derivation is not roundabout in that it contains only concepts which recur in the final result. (Gentzen 1934–1935, reprinted in Szabo 1969, p. 88)

Cut-free proofs thus are analytic in Bolzano's sense. Gentzen echoed Bolzano's aversion to the "atrocious detours" found in synthetic proofs. A cut-free proof, he wrote, "makes no detour" [*er macht keine Umwege*] (p. 69). One can even read Gentzen's metaphor of "constituent elements" as a reflection of Bolzano's idea that the *Abfolge* relation traces the reasons why facts obtain.

It is instructive, though, to observe the departures from and inversions of Bolzano's ideas that led Gentzen to his solution of the completeness problem. In Gentzen's systematization of logic, Bolzano's *Abfolge* relation, which had been an objective feature of the transcendent world, is replaced by logical rules immanent in mathematical practice. Because these rules are on display in actual mathematical reasoning, and not hidden away in the causal structure of the realm of propositions, they are observable. To discover them, Gentzen had only to devise the right conceptual grid to place on mathematical discourse. Bolzano's *Ableitbarkeit* relation, meanwhile, is simplified into a concrete definition of logical consequence. But whereas Bolzano conceived of the *Ableitbarkeit* relation as anthropomorphic rather than ontological,

---

[5] Szabo (1969) has "The S-formulae," preserving Gentzen's technical terminology.

and therefore faced the challenge of establishing a correspondence between one relation in the realm of objective dependencies and a second relation about derivability, Gentzen analyzed logical consequence directly into his proof system so that the analytic rules live alongside the synthetic CUT operation. There is no difference in kind between logical consequence and derivability, and completeness is no longer a matter of establishing a correspondence between the immanent and the transcendent. The logical rules of the sequent calculus are complete, both individually (by capturing the meanings of the logical particles they govern) and collectively, because proofs involving the CUT rule can be transformed into purely analytic proofs.

A surprising outcome of Gentzen's answer to the completeness question of Bolzano's *Wissenschaftlehre* (1937) is that it serves also as an elucidation of a central problem in his *Beyträge* (1810). In that earlier work, we saw Bolzano occupied with the question of whether or not "the propositions appearing in a scientific system" are always ordered so that the propositions with the more compound predicate follow those with the simpler predicate. In §27 he claimed that this is the case, but that one cannot also say that propositions with wider predicates follow those with the narrower predicates. The reason, he claimed, is that the syllogism rule disrupts the analyticity of proofs yet is for all we know unavoidable. This is because the analytical rules that he observed are not applicable to multiple "simple sentences," and it is possible that such sentences are found among the groundless, basic truths and the grounded, consequent truths alike. In §20 Bolzano observed that there is no way to reason one's way to simple propositions other than with the syllogism. So if there are such truths that are not basic, in Bolzano's sense, then there are truths that do not have purely analytical proofs. Obviously Gentzen's logical investigations do not touch on such ontological matters as which facts are basic. However, his cut-elimination theorem does corroborate Bolzano's claim, by showing that CUT is essential only when reasoning from axioms.[6] It is eliminable otherwise.[7]

The ontological dimension of Bolzano's thought cannot be easily evaluated. It elevated the *Abfolge* relation to a transcendent realm whose study proved too esoteric for scientific tools. This rendered the question of the adequacy of the *Ableitbarkeit* relation unanswerable and left vague the

---

[6] Significantly, CUT is Gentzen's slight modification of the rule Hertz called SYLLOGISM in the context of closely related research; see Hertz (1929), which Gentzen cites.

[7] In fact, an extension of Gentzen's work, due to Gasai Takeuti, shows that in sequent calculus proofs from axioms, only "anchored" CUTs are essential; "free" CUTs are eliminable. Anchored CUTs are instances of the CUT rule, at least one of whose premises is a descendant, in the proof-tree, of an axiom.

limits of normalization, preventing Bolzano from recognizing the subformula property of his analytic proofs. For these reasons, Gentzen's deflation of the transcendental aspect of logic appears to be just what was needed to put forward a scientifically respectable, and answerable, question of logical completeness. Indeed, Gentzen not only managed to pose the question of logical completeness in terms that admit precise solution, but he also managed to show that Bolzano's questions about completeness and analyticity of proofs are in the final analysis the same question. It is thus tempting to fault Bolzano's ontological preoccupations as the major obstacle in the way of modern logic. On the other hand, nothing like the question of logical completeness was posed in any form prior to Bolzano. The chasm that Gentzen (1932) filled, with his analysis of logical consequence and his *Hauptsatz* (1934–1935), is the one that Bolzano dug.

4 Gödel's answer

Unlike Gentzen, who closed the gap between *Abfolge* and *Ableitbarkeit* by projecting the latter relation into his proof system, today's prevailing conception of logical completeness preserves Bolzano's metaphysical divide between consequence and derivability. But whereas Bolzano, following Laplace and Lagrange, saw in the mechanical manipulation of signs a way of leaving intuition behind and tracking the objective dependencies among truths, this activity is today associated with derivability. *Ableitbarkeit*, which Bolzano considered anthropomorphically grounded, is recast in terms of set-theoretic semantics as the consequence relation that transcends effective processes with formal signs. One asks whether all the logical consequences of a set of sentences are in fact derivable from those same sentences. Bolzano's question is reversed.

   This reversal cannot be attributed without further ado to Gödel, who never described logical consequence as a semantic notion. Gödel defined "*logischen Folgens*" as "being formally provable in finitely many steps," and he nowhere departed from this usage. Indeed, Gödel did not formally define any semantic notions in his papers on logical completeness, appealing instead to intuitive ideas of satisfiability and validity that suffice for his purposes. Even Tarski, whose later work on logical consequence eventually coalesced into the definition familiar today, throughout his papers from the 1920s meant by the "consequences" of a set of sentences the larger set of sentences formally derivable from them.[8] Most remarkably, although today a simple rearrangement of ideas leads from what Gödel

---

[8] See for example the first five papers in Tarski (1956).

proved – that every universally valid, first-order formula is provable – to the statement of "strong completeness" – the derivability from a set of sentences of all their logical consequences – the first characterization of logical completeness in these terms did not appear until Robinson (1951), two decades after Gödel's thesis.[9] If Gödel neither invented nor articulated the modern syntax/semantics distinction, how did he manage to prove the theorem that made this nineteenth century relic respectable once again?

Although Gödel's work is well known, it is not usually read against the backdrop of the shifting configuration of ideas that surrounded it. Gödel, however, reflected on such matters in his late writing and considered the currents of thought relevant factors, not always positive, in the attainment of new results. In Gödel (1961) he characterized twentieth century work in the foundations of mathematics in terms of a priorism and empiricism, grouping the first of these notions together with theology, idealism, and metaphysics as being "on the right" of a dialectic, with empiricism, skepticism, and positivism on its left. "Now it is a familiar fact," he wrote, "even a platitude, that the development of philosophy since the Renaissance has by and large gone from right to left – not in a straight line, but with reverses, yet still, on the whole" (Gödel 1961, reprint in Gödel 1995, p. 375). However, he cautioned, "by its nature mathematics is very recalcitrant in the face of the *Zeitgeist*" (p. 379). Modern thought has managed to annex more and more questions into the domain of empirical investigation, often against deep-seated presuppositions that certain matters could not be so addressed. But even in the twentieth century, the opinion persisted that questions of mathematics are perhaps uniquely immune to this trend, are quintessentially a priori in nature, so that their reformulation in empirical terms will always be a mutilation of their original, properly mathematical sense. Referring specifically to the question of logical completeness, David Hilbert (1930, p. 140) remarked that "up till now we have come to the view that these rules suffice only through experiment [probieren]" and called for a mathematical proof that would provide more than evidence.[10]

[9] See Dawson (1993) for more on the late acknowledgement of the primacy of the semantic consequence relation and the role of Robinson's book. And notice that even Robinson's contribution was widely dismissed. In his review, Goodstein (1953) wrote "Only the first fifth of the book, which is devoted to an extension of Gödel's completeness theorem..., has any bearing on the foundations of mathematics, and the remaining four-fifths may be read without reference to this first part which could with advantage have been omitted."

[10] A passage in Hilbert and Ackermann (1928), evidently based on notes prepared by Bernays in 1917, makes the same point. Why is the question still unsolved? Because at present "It is only known purely empirically that this axiom system suffices for all applications."

Although he made his own affinity for rightward thinking clear and decried the "rabid" progression of thought "on the left" as overly pessimistic and a disservice to the nobility of human reason, Gödel acknowledged a certain propriety in the *Zeitgeist* and declared a partial allegiance to it: "Now one can of course by no means close one's eyes to the great advances which our time exhibits in many respects, and one can with a certain justice assert that these advances are due just to this leftward spirit in philosophy and world-view" (Gödel 1961, reprinted in Gödel 1995, p. 377). If we try to salvage mathematics by ignoring this fact, the result will not fit with the world we inhabit, i.e., the world as we view it. Gödel therefore advised a "workable combination" of empiricism and a priorism that "avoids both the death-defying leaps of idealism into a new metaphysics as well as the positivistic rejection of all metaphysics" (Gödel 1995, p. 387).

It has been suggested that this stage of Gödel's thought succeeded a more stridently "rightward" temperament that informed his youthful achievements. But Gödel's thesis can profitably be read as an exercise in the dialectic between the modern *Zeitgeist* and mathematical recalcitrance. What distinguished Gödel from other logicians was not solely his adherence to the classical sense of objective mathematical truth – other writers of his day resisted the encroachment of empiricism in mathematics. Gödel seems, rather, to have been uniquely attentive to both voices in this dialectic: he addressed the completeness question in a way only possible by heeding each.

This is evident especially in the reference Gödel made, in the introductory remarks to his thesis, to the conviction, shared by Hilbert, Poincaré, and other leading scientific voices, that ontological and semantic theories about existence and truth could be eliminated from scientific discourse, these concepts redefined nominalistically. Hilbert gave this idea a wide audience when, in his 1900 address to the International Congress of Mathematicians, he remarked:

> if it can be proved that the attributes assigned to the concept can never lead to a contradiction by the application of a finite number of logical processes, I say that the mathematical existence of the concept (for example, of a number or a function which satisfies certain conditions) is thereby proved. . . Indeed, when the proof for the compatibility of the axioms [of real numbers in analysis] shall be fully accomplished, the doubts which have been expressed occasionally as to the existence of the complete system of real numbers will become totally groundless. (Hilbert 1900a)

Hilbert's most detailed clarification of this "consistency implies existence" doctrine came in the course of an 1897–1902 correspondence with Gottlob

Frege about the foundations of geometry and axiomatic method. Frege had reported his view about the relationship between truth and consistency:

> I call axioms propositions that are true but are not proved because our knowledge of them flows from a source different from the logical source, a source which might be called spatial intuition. From the truth of the axioms it follows that they do not contradict each other. (Frege 1980, p. 39)

Recorded in these words is a vestige of Bolzano's *Grundlehre*: axioms are not a matter of convention, but are inherently unprovable facts. They are objectively true, the ground of the truth of other mathematical claims, and just as Bolzano knew that a proof that tracked the *Abfolgen* among propositions could not be circular, Frege cites the truth of each individual axiom as the reason that they cannot contradict one another. In his reply, Hilbert objected to each point:

> I found it very interesting to read this sentence in your letter, for as long as I have been thinking, writing, and lecturing on these things, I have been saying the exact opposite: if the arbitrarily given axioms do not contradict each other with all their consequences, then they are true and the things defined by the axioms exist. For me this is the criterion of truth and existence. (Frege 1980, p. 40)

Clearly Hilbert's intention to treat consistency as "the criterion of truth and existence" depends on the completeness of his logical rules. If those rules are not strong enough and sufficiently many, then a system could be free of contradiction for the irrelevant reason that the derivational apparatus is too meagre. In that case one would not want to conclude that the things defined by its axioms exist. Gödel observed, however, that the doctrine that mathematical existence and truth be reduced to the syntactic condition of consistency is an obstacle to a completeness proof. Logical completeness, he wrote (1929), "can easily be seen to be equivalent to the following: Every consistent axiom system consisting of only [first-order formulas] has a realization." He continued: "one might perhaps think that the existence of the notions introduced through an axiom system is to be defined outright by the consistency of the axioms and that, therefore, a proof [of the existence of a model, based on those axioms' formal consistency] has to be rejected out of hand." That proof requires one to demonstrate the link between consistency and existence – a task hard to appreciate by anyone who believes that mathematical existence can simply be defined in this manner. Thus the completeness proof that Hilbert sought would elude anyone who identified existence and consistency at a conceptual level. But Gödel objected further to Hilbert's idea that any consistent set of axioms implicitly defines a system of things as not merely requiring proof, but simply false. If axioms are supposed to fix their reference uniquely in the sense of being categorical, then they can only do this if they comprise a syntactically complete theory. For if a system, say first-order PA, is not syntactically complete, then there is some sentence S such that both PA+S and PA+¬S are consistent and, by the completeness theorem, satisfiable. In that case, PA has two distinct models and therefore cannot meaningfully be said to define any one structure. "This definition," Gödel concluded, "manifestly presupposes ...that every mathematical problem is solvable" (p. 61). (Gödel's most famous achievement is his demonstration that this presupposition is false, i.e., that many theories including PA are syntactically incomplete (Gödel 1931b).) Nominalism is self-refuting because its attempted redefinition of truth and existence depends on a theorem that falsifies those definitions.

Gödel did not present this refutation of the "left-leaning" attempt to supplant metaphysical truth with syntactic conditions as reason to return to Frege's traditional approach to questions of truth and consistency. Nor did he draw from his discovery a fully developed reversal of Bolzano's consequence/derivability scheme. He worked within the modern framework, preserving the syntactic notion of consequence, and merely investigated a tension internal to Hilbert's thought: between the doctrine that there is nothing more to truth than formal consistency and the idea that logical completeness must be proved. Thus the configuration of ideas familiar to logicians today, far from being necessary, is of rather recent vintage. Gödel set in motion a gradual reconfiguration of basic assumptions about logical form and content that began in the only possible place – where everyone around him was already standing. The semantic definition of consequence followed years later, the reformulation of completeness as a full reversal of Bolzano's question later still.

The scientific merits of reconcieving logical completeness are well known. If Gödel did not envision anything quite like modern set-theoretical semantics, he nevertheless showed that the realization of completeness as a correspondence of truth and proof had a further consequence: the compactness of first-order languages, what has perhaps been the single most applicable result in model theory. But Gödel's theorem has not only shaped proof theory and model theory in the decades since his work. The legitimacy of each as a branch of logic and the observed coincidences of their central concepts (e.g., theorems of Beth and Craig) rest on the fundamental correspondence that Gödel established.

Gödel's refusal to be swept away by nominalism played an essential role in his work on logical completeness. But even as he resisted the ideological currents surrounding formalism, Gödel was attentive to the formalisms themselves, the problems they raised, their merits and limitations. He wrote to Wang in 1972 that "Wittgenstein's negative attitude towards symbolic language is a step backward" (Wang 1996, p. 174). Gödel managed not to be drawn into either the nominalistic or the metaphysical conception of logic so deeply as to be unable to appreciate its rival. In his hands, in fact, the two conceptions were not rival: their hidden, partial affinity was the subject of his work.

5 A yellow rose

Gödel's vindication of the old metaphysical concept of mathematical truth against the encroachment of nominalism was doubly unlikely. Most obviously, as Gödel himself stressed, the momentum of positivism and empiricism that had accumulated since the Renaissance was, by the twentieth century, formidable. Scientific progress in the large, and modern logical research in particular, thrived on overturning a priori theories. The ontological character of Bolzano's thought, for example, was an obstacle to its scientific development. Even as it framed a clear and gripping completeness question, it stood in the way of several fundamental features of modern logic – the focus on formulas instead of propositions, the association of derivability with mechanizable processes, the disregard of factual truth – and ultimately, also, rendered its own central question unanswerable. Gödel noted that "the preceding rightward philosophy" was "excessive" and oriented in "the wrong direction," and, despite his ideological misgivings, he credited the stubborn sobriety of empiricism with correcting these faults (Gödel 1961, reprinted in Gödel 1995, p. 381).

Gödel's tenacity in the face of the *Zeitgeist* and his ability to inaugurate an entire research program based on a resistance against the prevailing scientific temperament is therefore striking. It is even more incredible in light of the conventionality of the formulation of completeness in terms of form/content correspondence. Students of logic today take Gödel's formulation to be *the* definition of logical completeness and see Gödel's theorem as inevitable. But a satisfactory reformulation of Bolzano's question did not require opposing the spirit of the age. The progression of ideas led very naturally to Gentzen's results, which show definitively that completeness can be dissociated from metaphysical truth and expressed as a property internal to a logical system. It is a marvel that a great scientific achievement of the twentieth century would involve reinstating the centrality of a priori, transcendent truth when the same question can be answered in the century's preferred terms.

Some readers might be tempted to fasten on the plurality of conceptions of logical completeness and the evitability of Gödel's accomplishment as evidence against the importance of his theorem and the correctness of the point of view that made it possible. If Gödel's conception of logic has viable rivals, then it is not the right way of looking at things in any objective sense. But closer consideration supports the opposite verdict. Yes, logic could well have developed around an alternative conception of completeness. The question Gödel answered was not "out there" – intelligible and pressing to everyone who worked in the field – but was rather an artifact of a particular way of looking at things that the current of modern thought made all but impossible. Gödel not only overcame these challenges, but he communicated his discovery with such clarity and force as to introduce a whole way of thinking so fundamental and useful that it has come to seem, in very short time, unavoidable. Gödel's theorem is monumental, not because it solved an eternal problem, but because it pins down a way of looking at logic that we might not otherwise have been acquainted with, yet could not today live without.

One cannot sail against the wind by ignoring it, only by reading it especially carefully. So too Gödel did not re-varnish the reputation of objective mathematical truth by turning his back on the tendencies of modern thought – he showed that empirical, formal logical investigations depend on this scorned fantasy.

> Then the revelation occurred: Marino saw the rose as Adam might have seen it in Paradise, and he thought that the rose was to be found in its own eternity and not in his words; and that we may mention or allude to a thing, but not express it; and that the tall, proud volumes casting a golden shadow in a corner were not—as his vanity had dreamed—a mirror of the world, but rather one thing more added to the world.[11]

[11] Jorge Luis Borges (1964), "A yellow rose."

PART III

Computability and analyticity

CHAPTER 6

Gödel's 1946 Princeton bicentennial lecture: an appreciation

Juliette Kennedy

1 Introduction

Two lines of thought run through what might be termed Gödel's program in the foundations of mathematics.[1] The large cardinal program, so-called, aims to isolate the "right" extension of the current ZFC axioms of set theory. Such an extension would decide independent statements such as the continuum hypothesis (CH henceforth), and would indeed eliminate *any* set-theoretic independence beyond so-called residual independence, i.e., that arising (unavoidably) from Gödel's incompleteness theorems. The motivation here, simply put, is to leave mathematics as it is: to apply the Law of Excluded Middle without restriction, from finitary mathematics to what Hilbert called the "ideal realm," including set theory.[2] The unrestricted application of the Law of Excluded Middle has been thought of as implementing a form of Platonism, by some; but such decidability was also a goal Gödel shared with Hilbert, if not necessarily with what is known as the Hilbert program (in proof theory).

Gödel's other program, that which occupies us in this chapter, has to do with considerations more purely logical in nature. This program is less known, though it preoccupied Gödel as much as the large cardinal program did. It complements the large cardinal program, is interleaved with it, and in some sense subsumes it. The program[3] is based on a collection of ideas associated with the term "informal rigor," to use Kreisel's expression for this theme in Gödel's writings; and though the mathematical prescription considered here differs with the program Kreisel (1967) outlines[4] the aim is

[1] This chapter is a companion to section 5 of Kennedy (2013), which also treats Gödel (1946). The treatment given here is philosophical rather than programmatic, and focuses entirely on Gödel's philosophical work rather than on current mathematical practice.
[2] Including, in particular, those set-theoretic propositions now known to be independent, such as the CH.
[3] The word "program" may indicate a more explicit set of commitments than what appears in Gödel's writings (in this connection) at first glance. We hope to establish that the term is warranted.
[4] That program has to do with the pursuit of independence results.

similar: to delineate the presence and the reach of logical methods, and then attempt to recover a decision procedure for mathematics that lies, roughly speaking, beyond the reach of such methods; to formulate a set of principles, even a system, which is exact enough to encompass all of mathematical reasoning, but informal. This is simply the "old-fashioned idea," as Kreisel puts it, "that the intuitive notions are *significant*."[5]

Gödel's 1946 Princeton Bicentennial Lecture[6] is not perhaps the origin of this line of thought,[7] but in our view it contains the most explicit and exact formulation of it. In fact, seen from a "formalism free" perspective, one finds here the seeds of a new foundational program, and within foundations a newly assessed role of formalization.[8]

2 Gödel's remarks to the 1946 Princeton bicentennial conference

Gödel begins the lecture with three *epistemological* notions, as he calls them: computability, provability and definability. What does Gödel mean here by the term "epistemological"? Each of these terms represents activities of the mathematician; also each generates a theorem about paradoxicality: the unsolvability of the halting problem, the Incompleteness Theorems and paradoxes arising from the attempt to define definability, respectively. For example, "the least undefinable ordinal" defines an ordinal, paradoxically. One guesses that Gödel uses the term "epistemological" in opposition to the term "ontological," the latter category subsuming concepts such as meaning or truth. As we will see below, by the 1970s Gödel would partition the mathematical field by means of the categories "subjective" and "objective" – Gödel's transfer of the epistemic/ontologic distinction, if you like, at least relative to mathematics.

In his introduction to Gödel's essay "Russell's mathematical logic" in volume II of Gödel's collected works (Gödel 1990), Charles Parsons made the important observation that Gödel's realism, "which has excited much critical comment," expresses itself *programmatically*, "in the context of concrete problems and as motivating mathematical research programs"

---

[5] The passage occurs in the opening lines of Kreisel (1967).

[6] The lecture was delivered in Princeton in 1946. It appeared in print for the first time in Davis (1965), and appeared subsequently in Gödel (1990), together with an introduction by Charles Parsons.

[7] For example, the (related) concept of absolute provability occupied Gödel already in his 1929 thesis. See Kennedy (2011).

[8] The concept called formalism freeness in Kennedy (2013) generalizes the phenomenon Gödel (1946) refers to as "formalism independence." One can think of formalism freeness as the simple preference for semantic methods, that is, methods which do not involve or, more precisely, *do not require* the specification of a logic. But the phenomenon manifests itself in other ways as well.

(Gödel 1990, p. 107). Parsons' remarks that Gödel's method, namely designing mathematical research programs in order to implement his philosophical views – research programs which, when successful, represent the "cash value" (Gödel 1990, p. 107) of those philosophical views – reached a high point in Gödel's 1947/64 essay, "What is Cantor's continuum problem?" (Gödel 1947). But one could equally well see a high point as having been achieved in Gödel's 1946 lecture, with its isolation of ordinal definability, a concept whose "cash value" has turned out to be very high, ordinal definability and also hereditary ordinal definability[9] being utterly fundamental to set-theoretic practice nowadays.

Accordingly the goal in Gödel (1946) is to obtain formulations of each of the epistemological notions, computability, definability and provability, which are "absolute." By "absolute" Gödel means the following: one expresses the notion in some suitably chosen formalism, and then if this is done in an absolute fashion, no new objects satisfying the defining expression will be definable in any (sensibly chosen, i.e., standard) extension of the ground theory. The challenge of course is to do this in such a way as to avoid paradox. In Gödel's view this has been already achieved with the notion of computability, via Turing's analysis, and indeed his suggestion here is simply to transfer the Turing analysis to two remaining epistemological notions, provability and definability.

As we will see below, Gödel's notion of absoluteness, as relativized to definability and provability, is based on nondiagonalizability. What this means in the context of ordinal definability is the following: if one passes to the "next language," i.e., a language obtained by adding a truth predicate for statements about ordinal definable sets, one obtains no new ordinal definable sets.[10] On the other hand, obtaining nondiagonalizability in the case of provability is problematic, and indeed remains so to this day.

Mathematically speaking, the address consists of conjectures and remarks, with no proofs given for the latter. For example, Gödel's assertion mentioned above, that ordinal definability is definable, depends on the Lévy Reflection Principle – but the proof of the Principle appeared in print only in 1960.[11]

---

[9] For the definitions of these terms see below.

[10] The class of ordinal definable sets, denoted OD, are those sets which are definable by a formula of set theory with finitely many ordinal parameters. HOD is the class of hereditarily ordinal definable sets, i.e., sets whose transitive closure is ordinal definable. In the lecture, Gödel only mentions OD; but we take him to be referring to HOD in most cases.

[11] The Lévy Reflection Principle states that for every n there are arbitrarily large ordinals α such that $V_\alpha \prec_n V$, i.e., such that V_α is an n-elementary submodel of V, see Lévy (1960–1961).

Another remark of Gödel's is that the Axiom of Constructibility is independent of the ZFC axioms of set theory. About this question, J. R. Schoenfield wrote the following in 1959:[12]

> It is natural to ask if the Axiom of Constructibility follows from the Axiom of Choice and the Generalized Continuum Hypothesis. A difficulty in answering this question is that the independence of the Axiom of Constructibility from the axioms of set theory... has not been settled. (Schoenfield 1959)

Gödel's 1967 letter to Wolfgang Rautenberg (Gödel 2003b, p. 183) contains a claim to the contrary. Rautenberg had written to Gödel asking about a claim Mostowski had made, that "it has been known since 1938 that Gödel has a proof of the independence of these hypotheses."[13] Gödel replied that Mostowski's claim was incorrect; however he had obtained certain partial results, "namely proofs for the independence of the axiom of constructibility and the axiom of choice in type theory." Gödel added that it was doubtful that his method, which he says, interestingly, was related to Scott and Solovay's Boolean-valued models "rather than Cohen forcing," "would be at all sufficient for such a result."[14]

As for the remaining mathematical conjectures of Gödel's in the Lecture, they involve the question whether HOD is a model of set theory satisfying the Axiom of Choice, a result that would give a simpler consistency proof of that axiom. This was eventually proved by Myhill and Scott (1971). Finally, Gödel predicts that the proof of the Axiom of Choice in HOD will not extend to a proof of the CH, and in fact the failure of the CH was shown to be consistent with $V = \text{HOD}$ in 1968 (see McAloon 1970–1971).

2.1   Grounding effectivity

Gödel opens the lecture with the concept of computation, remarking that the concept can be given, in his words, a "formalism independent" definition:

> Tarski has stressed in his lecture the great importance (and I think justly) of the concept of general recursiveness (or Turing computability). It seems to me that this importance is largely due to the fact that with this concept one

---

[12] For the definition of the Axiom of Constructibility see below.

[13] Namely the continuum hypothesis and the Axiom of Constructibility, see Mostowski (1964).

[14] That is, the continuum hypothesis. For further background on this issue, see Gödel's August 1966 correspondence with Church in Gödel (2003a). See also Parsons's introductory note to the Rautenberg correspondence in Gödel (2003b), pp. 179–180.

has succeeded in giving an absolute definition of an interesting epistemological notion, i.e. one not depending on the formalism chosen. (Gödel 1990, p. 150)[15]

Gödel is referring here to two things. First there is the phenomenon Gandy (1988) calls "confluence," namely the fact that all of the known mathematical notions of computability, i.e., the Gödel–Herbrand–Kleene definition (1936), Church's λ-definable functions (1936), Gödel–Kleene μ-recursive functions (1936), Turing machines (1936), Post (1943) systems, or Markov (1951) algorithms (see Davis 1958), define the same class of functions. Second, a sense of formalism independence emerges in the paper, having to do with *absoluteness* in the sense defined above, namely the stability of a concept belonging to a theory, relative to suitable extensions (of the theory).[16]

In a footnote appended to this first sentence in 1965,[17] Gödel offers the following clarification (italics added):[18]

> To be more precise, a function of integers is computable in *any* formal system containing arithmetic if and only if it is computable in arithmetic, where a function f is called computable in S if there is in S a computable term representing f.

Gödel is just referring to the fact that any computable (i.e., partial recursive) function representable by a term in a formalism *extending* arithmetic, is already representable in Peano arithmetic, by the very same term.

The 1965 clarification was needed, as Gödel had actually conflated two essentially different concepts in the opening lines of the 1946 lecture. "Formalism independence," taken in the sense of "confluence," applies to a broad range of conceptually distinct notions of computability, and is established by proving the equivalence of the corresponding encodings of these various notions in a suitably chosen metatheory.[19] One might also think of a secondary notion of confluence, operating at the informal level, for example, recognizing that an equivalence holds between Gödel–Herbrand

---

[15] This is the first sentence of the lecture.
[16] As Parsons puts it in his introduction to Gödel (1946), in Gödel (1990), Gödel is referring to "the absence of the sort of relativity to a given language that leads to stratification of the notion such as (in the case of definability in a formalized language) into definability in languages of greater and greater expressive power." The stratification is "driven by diagonal arguments."
[17] To the version of the lecture published in Davis (2004).
[18] This is Sieg's (2006a) broadened absoluteness claim.
[19] A rather weak metatheory is required for this. One codes the various calculi, e.g., Turing machines, or Herbrand–Gödel systems of equations, and then one proves their equivalence in the metatheory, where the proofs are usually facilitated by the Kleene T-predicate.

computability and (unformalized) Turing machine computability, but taking the equivalence as proved in the metatheory as evidence.

However, in the sense given in Gödel's 1946 lecture and also subsequently in the 1965 footnote, absoluteness is a property of a particular class of formal systems, as noted by Sieg (2006a). A restricted form of confluence, absoluteness applies "piecewise," so to speak, to particular notions of computability, which are absolute relative to (a particular class of) their extensions. Thinking in terms of explanatory priority, the existence of the sharp, informal notion "performable by a Turing machine" explains, i.e., is the reason behind, confluence; but then confluence in turn witnesses the robustness of the sharp informal notion. Absoluteness then comes as a corollary – for free, so to speak, being a particular kind of confluence.

Back to the lecture. Gödel contrasts the absoluteness of the concept of computability with the failure to secure the same in the cases of provability and definability:

> In all other cases treated previously, such as demonstrability of definability, one has been able only to define them relative to a given language, and for each individual language it is clear that the one thus obtained is not the one looked for. For the concept of computability, however, although it is merely a special kind of demonstrability or definability, the situation is different. By a kind of miracle it is not necessary to distinguish orders, and the diagonal procedure does not lead outside the defined notion.

As is well known, Turing's analysis of the notion of "formal system" actually convinced Gödel of the generality of the First Incompleteness Theorem in the late 1930s. We now recount this history, as it bears on our interpretation of Gödel's 1946 lecture. A precise notion of formal system was needed for settling the question, taken up by Gödel himself in his 1931 paper, whether the Incompleteness Theorems are sufficiently general, that is, whether they apply to any formal system containing arithmetic, and not just *Principia* and systems related to it. Gödel was careful to say at the end of his 1931 paper that this had not been shown, and indeed the issue lingered for some time after the Incompleteness Theorems had been published.[20] But Turing's analysis lays this doubt to rest for Gödel.[21]

---

[20] See, for example, Sieg (2006b). Also in this connection, the recursion theorist Robert A. di Paola once related the following story to the author: some time in the early 1970s di Paola called Gödel at his home and inquired about the generality of the Incompleteness Theorems, to which Gödel replied, "They are *perfectly* general!!"

[21] Kreisel (1967) takes a similar line.

In a postscript Gödel added on the occasion of the reprinting of the lecture in Davis (1965), he writes:

> In consequence of later advances, in particular of the fact that, due to A. M. Turing's work, a precise and unquestionably adequate definition of the general concept of formal system can now be given, the existence of undecidable arithmetical propositions and the nondemonstrability of the consistency of a system in the same system can now be proved rigorously for every consistent formal system containing a certain amount of finitary number theory. (Gödel 1986, p. 369)

The generality issue arose originally because the scope of the Incompleteness Theorems, that is to say the extent of the phenomenon of incompleteness relative to the arithmetic theories known at the time, was not clear from the 1931 paper. But, Gödel continues:

> Turing's work gives an analysis of the concept of "mechanical procedure" (alias algorithm or computation procedure or "finite combinatorial procedure"). This concept is shown to be equivalent with that of a "Turing machine." A formal system can simply be defined to be any mechanical procedure for producing formulas, called provable formulas. For any formal system in this sense there exists one in the [usual] sense that has the same provable formulas (and likewise vice versa)... (Gödel 1986, p. 369)

Gödel's suggestion that the Turing analysis resolved the generality issue associated with the incompleteness theorems by giving a complete clarification of the notion "formal system," is based on a train of thought which is given only partially in Gödel's writings. We now attempt to flesh out this train of thought, and in order to do this some background is required. During the period preceding Turing's landmark paper (Turing 1936), particularly during the years 1934–1935, both Church and Gödel analyzed the concept of "effectively computable function" in terms of calculability in a suitably chosen logic (or class of logics). For Church, the condition placed on the logic in question was (essentially) that the theorems of the formal system should be recursively enumerable. As Sieg (2006b) relates, Church argues for this recursive enumerability on the basis of the conditions given in Gödel's 1934 Princeton lectures:[22]

> These conditions, Church remarks in footnote 21, are substantially those from Gödel's Princeton Lectures for a formal mathematical system: (*i*) each rule must be an effectively calculable operation, and (*ii*) the set of rules and axioms (if infinite) must be effectively enumerable. Church supposes that

---

[22] For Gödel's 1934 lectures see Gödel (1986, p. 346).

these conditions can be interpreted to mean that, via a suitable Gödel numbering for the expressions of the logic, (i_c) each rule must be a recursive operation, (ii_c) the set of rules and axioms (if infinite) must be recursively enumerable, and (iii_c) the relation between a positive integer and the expression which stands for it must be recursive. The theorem predicate is thus indeed recursively enumerable, but the crucial interpretative step is not argued for at all.

Sieg observes that Church's analysis of the notion "effective," even as sharpened by Gödel's conditions i_c–iii_c, is nevertheless *semi-circular* – one defines effectivity in terms of calculability in a logic, and this logic is itself to be effective.[23] Hilbert and Bernays sharpen Gödel's conditions still further, with the crucial requirement that the proof predicate of the logic is primitive recursive (see, for example, Sieg 2006a). Strictly speaking, this does not solve the circularity problem either.[24] In fact, any analysis of effectivity given in terms of calculability in a logic which is itself effectively given will be subject to the charge of semi-circularity – at least at first glance.

In fact, instead of a semi-circularity, the Church and Gödel analysis during the period 1934–1935, as well as the analysis of Hilbert and Bernays, rather appear to give rise to an *infinite regress*: effectivity, thought of informally, is explained via a logic which is itself given effectively (e.g., in the sense of Gödel's conditions i_c–iii_c). But how then to explain this second sense of effectivity, i.e., as applied to the logic? One must introduce a third formalism, by means of which the effectivity of the logic is to be analyzed. And so forth.

We now return to Gödel (1946). Sieg questions Gödel's broadened absoluteness claim, i.e., Gödel's assertion that "a function of integers is computable in *any* formal system containing arithmetic if and only if it is computable in arithmetic," (italics added) as according to Sieg, "Obviously, one has to exploit in some way the formal character of the system, for example, by making the assumption of Church or by imposing the recursiveness conditions of Hilbert and Bernays. Either way of proceeding reveals the relativity of Gödel's absoluteness."[25] Strictly speaking, "the correctness of the absoluteness claim depends on the fact that one is dealing

[23] This is Church's "step by step" argument; see Gandy (1988) and also Sieg (2006b).

[24] Of course, one could simply declare that a particular formalism has an effective proof predicate. But then issues of intensional adequacy arise. For the literature on this see Detlefsen (1979), Feferman (1960–1961), Pudlak (1996) and Frank (2009).

[25] Sieg (2006b). Sieg does note that Gödel's earlier, more restricted absoluteness claims do mesh with Church's step by step argument.

with appropriately restricted formalisms" (Sieg 2006b). Even worse, as Sieg notes here, Gödel seems to explain the notion of formal system in terms of computability in appropriate formalisms – another circularity.

It is exactly here, we suggest, that the nature of Turing's analysis becomes relevant and the various circularities, or alternatively an infinite regress, prevented. The crucial point is this: rather than calculability in a logic, Turing analyzes effectivity in an *informal but entirely convincing* manner, via the concept of a Turing machine – a formalism free but fully exact mathematical notion consisting of a machine model, i.e., a tape scanned by a reader, together with a set of simple instructions in the form of quadruples, which, if carried out in the right way, reduce the computation of, for example, a λ-definable function, to child's play.

Turing's informal machine model *grounds* the notion of "human effective computability." Moreover this grounding, this fitting together of two seemingly distinct concepts, human calculability "following a fixed routine" (quoting Gandy 1988) and Turing calculability, a fitting together so strongly convincing as to be called a theorem by Gandy,[26] takes place at the informal level – and indeed *it has to be this way*. We note the point once again: Gandy states the theorem as establishing a correspondence between the notion of human calculability following a fixed routine, a necessarily informal notion, and Turing machine calculability in the informal sense of the latter. What is the role of confluence? That the Turing machines, suitably formalized, are provably equivalent to the other known notions of provability, suitably formalized, establishes confluence one level up, so to speak. This gives further evidence that the right class of functions, the "sharp concept" (as Gödel called it) has been isolated. But as was noted by Gödel in the 1930s and by Gandy subsequently, heuristically speaking, "confluence" is weaker than "grounding."[27] This is why Gödel, even so that he may have been among the first to formulate Church's thesis,[28] was reluctant to endorse the thesis, insofar as it relied solely on confluence, prior to Turing's work.

Turing's analysis sidesteps the infinite regress problem, in that grounding the notion "humanly computable" does not now require the specification of

---

[26] Church, on the other hand, refers to the correspondence as obviating the need for a theorem, see Gandy (1988).

[27] Both Gödel and Gandy use different language, of course (Gandy 1988).

[28] In suggesting that "an (all-embracing) characterization of computable function might be obtainable" (Gandy 1988, p. 72). Church's thesis, as originally suggested by Church in 1934, identified effective calculability with λ-calculability. In its present form Church's thesis, or more correctly the Church–Turing thesis, identifies human effective calculability with λ-calculability together with all notions of computability equivalent to it.

a logic or formal system.[29] It also convinces Gödel of the importance of confluence, all previous notions of computability being now provably equivalent to an entirely convincing model of human computability, i.e., the Turing machine.[30] The robustness of the notion of effectively computable function now being fully established, so then too is the robustness of the notion of "formal system," the latter being just a form of calculability.

We emphasize that the nature of Turing's analysis was decisive. As Gödel would later put to Hao Wang, Turing's model of human calculability is, in some sense, perfect:

> The resulting definition of the concept of mechanical by the sharp concept of "performable by a Turing machine" is both correct and unique ... *Moreover it is absolutely impossible that anybody who understands the question and knows Turing's definition should decide for a different concept.* (Wang 1996, p. 203, italics added)

> The sharp concept is there all along, only we did not perceive it clearly at first. This is similar to our perception of an animal far away and then nearby. We had not perceived the sharp concept of mechanical procedure sharply before Turing, who brought us to the right perspective. (Wang 1996, p. 205)

It is ironic that the "epistemic work," so to speak, the sharpening of intuition which is usually brought out by means of formalization of those very intuitions, should rather be brought out by the informal notion – the informal analysis in question here being actually more convincing than the various "formalism entangled" analyses which were prevalent prior to 1936.

Gandy sees Turing's isolation from the logical milieu of Princeton as the key to his discoveries:

> It is almost true to say that Turing succeeded in his analysis because he was not familiar with the work of others... The bare hands, do-it-yourself approach does lead to clumsiness and error. But the way in which he uses concrete objects such as exercise book and printer's ink to illustrate and control the argument is typical of his insight and originality. Let us praise the uncluttered mind. (Gandy 1988, p. 83)

The "work of others" that Gandy refers to here was driven by the prevailing, possibly obscurantist, idea of "calculability in a logic":

[29] By a formalism, or a *logic*, we mean a combination of a list of symbols, commonly called a signature, or vocabulary, rules for building terms and formulas, a list of axioms, a list of rules of proof, and finally a definition of the associated semantics.

[30] As was noted, Gödel was not convinced by confluence prior to the Turing analysis, see also Sieg (2006b).

Soon after, Gödel, in his lectures at Princeton. . . proposed the characterization of (what Kleene was to call) general recursive functions. The key idea
here is that of computability in a formal system; it goes back to Hilbert's
1905 and had been developed and exploited by Gödel. . . What was lacking
was a thoroughly cogent generalization of "formal system."

All the work described in Sections 14.3–14.9[31] was based on the mathematical and
logical[32] (and *not* on the computational) experience of the time. What Turing
did, by his analysis of the processes and limitations of calculations of human
beings, was to clear away, with a single stroke of his broom, this dependence
on contemporary experience, and produce a characterization which—within
clearly perceived limits—will stand for all time. (Gandy 1988, p. 101)

For Sieg, what was groundbreaking about Turing's analysis, its philosophically "genuinely distinctive character," as Sieg refers to it, what he moreover
claims both Church and Gödel failed to see in it, was that Turing solved the
Entscheidungsproblem in a way that did full justice to the "normative
demand for intersubjectivity between humans" (Sieg 2006a).

We are suggesting then a different scenario: Gödel's appropriation of the
Turing analysis in connection with the generality issue and in connection
with the approach to definability and provability given in his 1946 lecture
lends itself to the interpretation that by 1946 Gödel had actually dropped
the idea of cashing out the notion of effectivity in terms of calculability in
a logic. What is needed is the concept "computable by a Turing machine,"
an *entirely convincing yet informal* concept.

2.2　Entanglement

Before turning to Gödel's remarks on absolute provability, we allow ourselves a digression. We have offered a two-tiered analysis of the emergence of
the mathematical notion of effectivity. The elements of our analysis are the
following: first a *grounding* on the informal level of the informal notion of
human calculability via Turing's formalism free concept of Turing machine;
followed by *confluence*, i.e., the provable equivalence, in the metatheory, of
the formalized analogs of all the known concepts of computability.

The scenario requires a clear division between syntax, on the one hand,
and semantics on the other. We have argued elsewhere (Kennedy 2013)
that syntax and semantics are in some sense entangled, and indeed this is
doubly true here. If one defines a formalism, or alternatively a *logic*, as we

---

[31] This refers to the work on computability of Church and others prior to Turing's 1936 paper.
[32] Emphasis added.

have done in this chapter, i.e., as a combination of a list of symbols, commonly called a signature, or vocabulary, rules for building terms and formulas, a list of axioms, a list of rules of proof, and finally a definition of the associated semantics, then by a slight shift of perspective *each of the informal notions of computability we have considered so far can equally well be regarded as formal calculi in this sense* – individually. That is, each of these constitutes an individual, self-standing formalism with its own syntax and rules of proof and so forth.[33] What is notable in the case of the logical calculi mentioned here, of course, is that the associated semantics give the same class of functions in each case.

2.3   Absolute provability

We now turn to Gödel's suggestion that the Turing analysis of computability should be transferred to the other two epistemological notions, namely provability and definability:

> This, I think, should encourage one to expect the same thing to be possible also in other cases (such as demonstrability or definability). It is true that for these other cases there exist certain negative results, such as the incompleteness of every formalism... But close examination shows that these results do not make a definition of the absolute notions concerned impossible under all circumstances, but only exclude certain ways of defining them, or at least, that certain very closely related concepts may be definable in an absolute sense. (Gödel 1990, p. 150)

Gödel's preoccupation with the concept of informal provability is evident already in the first lines he ever wrote (as a logician), i.e., in the initial remarks of his 1929 thesis (see Kennedy 2011). Informal provability appears subsequently in Gödel's note [1933f] (Gödel 1986, p. 300), in which he offers an axiomatization of modal propositional logic (now known as $S4$) by means of the unary operator "*beweisbar*," denoted B, meaning "provable by any correct means" (introduction by Troelstra, Gödel 1986, p. 296) but notably not interpretable as "provable in a certain formal system S" (Gödel 1986, p. 301). And absolute provability was a major concern of Gödel's Gibbs Lecture, delivered in 1951 (Gödel 1995, p. 304). Gödel's remarks in the Gibbs Lecture do indicate a marked change in methodology. As was mentioned,

---

[33] On "seeing an aspect," see also Floyd's "On being surprised: Wittgenstein on aspect perception, logic and mathematics," in Krebs and Day (2010). Indeed, depending on one's perspective, most mathematical constructions given in natural language can be regarded as generating a formalism. The theory of groups, for example, given by a set of axioms stated in natural language, can easily be seen as generating a formal theory with an exact vocabulary and rules of proof and so forth.

much of Gödel's philosophical work prior to 1951 involved somewhat detailed mathematical implementations – his work on rotating universes during the years 1946–1949 being a prime example – whereas the argumentation in the Gibbs Lecture is almost wholly philosophical, focusing on the Incompleteness Theorems but involving no new theorems per se.

Gödel's 1946 lecture is still very typical of the programmatic approach which characterized his work throughout the 1940s:

> Let us consider, e.g. the concept of demonstrability. It is well known that, in whichever way you make it precise by means of a formalism, the contemplation of this very formalism gives rise to new axioms which are exactly as evident and justified as those with which you started, and that this process of extension can be extended into the transfinite. So there cannot exist any formalism which would embrace all these steps; but this does not exclude that all these steps ... could be described and collected in some nonconstructive way. In set theory, e.g., the successive extensions can be most conveniently be represented by stronger and stronger axioms of infinity. It is certainly impossible to give a combinational and decidable characterization of what an axiom of infinity is; but there might exist, e.g., a characterization of the following sort: An axiom of infinity is a proposition which has a certain (decidable) formal structure and which in addition is true. Such a concept of demonstrability might have the required closure property, i.e., the following could be true: Any proof for a set-theoretic axiom in the next higher system above set theory (i.e. any proof involving the concept of truth which I just used) is replaceable by a proof from such an axiom of infinity. It is not impossible that for such a concept of demonstrability some completeness theorem would hold which would say that every proposition expressible in set theory is decidable from the present [ZFC] axioms plus some true assertion about the largeness of the universe of all sets. (Gödel 1990, p. 151)

What Gödel is suggesting here is that a hierarchy of large cardinal assumptions should replace the hierarchy of formal systems generated by, for example, adding consistency statements to set theory, i.e., passing from ZFC to ZFC + Con(ZFC) and then iterating this, or alternatively by adding a satisfaction predicate for the language of set theory, then considering set theory in the extended language, and iterating this.

Gödel's idea that the transfinite hierarchy assists decidability in set theory eventually became part of Gödel's program for large cardinals, a program laid out in its fullest form in his 1947 "What is Cantor's continuum problem?"[34] and tied to the project of finding a single so-called intended model for set theory.

---

[34] See Gödel (1947). In fact very few large cardinals had been discovered in 1946.

Does Gödel's conjectured completeness theorem: "every proposition expressible in set theory is decidable from the present [ZFC] axioms plus some true assertion about the largeness of the universe of all sets" ground the natural notion of consequence analogous to the way that Turing's analysis grounds human calculability? It is perhaps relevant to note that the notion of cardinal number is itself "non diagonalizable": taking the supremum of a set of cardinals, for example, does not lead outside of the notion of cardinality. Also, large cardinal concepts are generally given semantically, i.e., in terms of embeddings of the universe into an inner model, where preservation of different properties of the embeddings – for example, conditions imposed on critical points – yield different cardinals.

In this connection it is perhaps interesting to take note of "the road not taken" by Gödel here. We saw previously that for Gödel formal provability is a special kind of computability. But what about informal provability? Is this not also a form of computation (in the sense of Section 2)? In fact Gödel's view of mathematical practice would not permit subsuming informal consequence under that of Turing computability. Gödel's view of the practice – which is rather firmly centered on consequence, after all – is that it involves, ineliminably, "the mind and abstract entities"; involves, ineliminably, *meaning*. As Gödel would later write to Leon Rappaport:

> My theorems only show that the mechanization of mathematics, i.e., the elimination of the mind and of abstract entities, is impossible, if one wants to have a satisfactory foundation and system of mathematics. (Gödel 2003b, pp. 176–177)

And as he would remark in the opening lines of his *Dialectica* paper:

> ... by abstract (or nonintuitive) notions we must understand those that are essentially of second or higher order, that is, notions that do not involve properties or relations of *concrete objects* (for example, combinations of signs), but that relate to *mental constructs*, (for example, proofs, meaningful statements, and so on); and in the proofs we make use of insights, into those mental constructs, that spring not from the combinatorial (spatiotemporal) properties of the sign combinations representing the proofs, but only from their *meaning*. (Gödel 1990, p. 241)

Gödel's basic picture, then is two tiered: the mathematician is, at times, a calculator on finite strings, in the sense of Turing; but the mathematician is also involved in a fully contentual discourse through her access to infinitary concepts.

2.3.1   Prescience?

The mathematician Michael Harris asks the question: "How can we talk to one another, or to ourselves, about the mathematics we were born too soon to understand?" (Harris, to appear). That is, Gödel's infinitary completeness theorem as stated in his 1946 lecture may actually have been witnessed, to a degree, by contemporary developments. In particular, the following result of Woodin might be viewed as a partial realization of the suggestion to replace logical hierarchies by infinitary principles: in the presence of large cardinals,[35] the Σ_1^2 theory of real numbers, i.e., existential statements about sets of reals, is (set) forcing immune in the sense that their truth cannot be changed by forcing – a strong form of absoluteness.[36] Another result of this kind is due to Woodin and says about the structure $L(\mathbb{R})$, the constructible closure of the reals, that its first-order theory is (set) forcing absolute in the presence of large cardinals (i.e., a proper class of Woodin cardinals, see Woodin (1988)).

If the Woodin results provide a strong form of absoluteness, is the concept of provability at issue in these results formalism independent, in Gödel's sense? An ω-proof, a proof concept which occurs in Woodin's later work in connection with generic absoluteness (Woodin 2010a), is just a universally Baire set of reals. Replacing the concept of a proof by simply a set of reals has an appearance of formalism freeness, as we have called it, somewhat reminiscent of the idea of replacing a formula by a set invariant under automorphisms in the AEC context.[37]

ω-provability is a vast topic but limitations of space preclude further discussion of it. As an alternative to ω-provability, an obvious grounding of the notion of consequence would seem to be given through the notion of, simply, semantic consequence. Gödel does not suggest anything of the kind in the lecture; and although his remarks on absolute provability here are far from perfunctory, judging from the space Gödel devoted to them, his mind was clearly on definability, to which we now turn.

2.4   Absolute definability

Gödel remarks about definability, that he can give us "somewhat more definite suggestions." The most important of these in connection with definability, mathematically speaking, is ordinal definability, a notion

---

[35] A proper class of Woodin cardinals is needed.
[36] The result requires the continuum hypothesis, see Woodin (2010a).
[37] For more on formalism freeness in the AEC context see Kennedy (2013).

which turned out to be of great importance for set theory and which is presented here for the first time.[38] The idea is to take the ordinals as already given and then define sets by means of the language of set theory as usual, but with finitely many ordinals as parameters.

> Here you also have, corresponding to the transfinite hierarchy of formal systems, a transfinite hierarchy of concepts of definability. Again it is not possible to collect together all these languages in one, as long as you have a finitistic concept of language, i.e as long as you require that a language must have a finite number of primitive terms. But, if you drop this condition, it does become possible ... by means of a language which has as many primitive terms as you wish to consider steps in this hierarchy of languages, i.e. as many as there are ordinal numbers. The simplest way of doing it is to take the ordinals themselves as primitive terms. So one is led to the concept of definability in terms of ordinals... This concept should, I think be investigated. (Gödel 1990, p. 151)

Ordinal definability is then nondiagonalizable – if one passes to the "next language," i.e., one obtained by adding a truth predicate for statements about ordinal definable sets, one obtains no new ordinal definable sets – and thus is absolute in Gödel's sense. But it is important to note that although the property of being an ordinal is very strongly absolute in the standard set-theoretical sense, ordinal definability and also hereditary ordinal definability are not absolute in the standard set-theoretical sense. Whether a set is ordinal definable or not can be changed by forcing – a fact of which Gödel would have been completely unaware in 1946.

In any case, in Gödel's sense of the term, ordinal definability is "absolute." As we saw, the Turing analysis, which Gödel wishes to transfer to the definability context, goes well beyond absoluteness. It involves two elements: confluence, and what we called "grounding." As for the latter, such a goal seems utterly out of reach in connection with definability, a concept which seems to be very entangled with the notion of a formalism. One uses a signature and formation rules to build up well-formed formulas, and then the definable sets are simply declared to be those sets that are definable via that stock of formulas. But as in the 1930s in connection with computability, definability in a logical calculus does not necessarily lead to the correct analysis. What Gödel seeks here is a grounding of the epistemological notion "comprehensibility by our mind" (see Gödel 1990, p. 152).

As the mathematical witness, so to speak, of "comprehensibility by our mind," ordinal definability might appear to be a poor choice, just from

[38] Post rediscovered ordinal definability in 1952. See Parsons, in Gödel (1990), p. 148.

cardinality considerations alone. The ordinals, being the "spine" of the cumulative hierarchy, are of course a proper class; whereas, as Gödel remarks, it is very plausible that "all things conceivable by us are denumerable, even if you disregard the question of expressibility in some language" (Gödel 1990, p. 152). Gödel's answer to the objection is that ordinals are "formed according to a law," and thus they bequeath their "lawlikeness" to the sets constructed from them, namely the ordinal definable sets.

Gödel turns next to constructibility as a possible candidate for the notion of absolute definability. The presentation of the constructible hierarchy, denoted L, for "lawlike,"[39] was given by Gödel in 1939 in his monograph on the consistency of the continuum hypothesis (Gödel 1940, reprinted in Gödel 1990). The constructible sets are for a model of the ZFC axioms of set theory together with the CH, and are defined as follows:

$$L_{\alpha+1} = \{X \subseteq L_\alpha : X \text{ is definable}$$
$$\text{over } (L_\alpha, \in) \text{ with parameters}\}$$
$$L_\nu = \cup_{\alpha < \nu} L_\alpha \text{ for limit } \nu \qquad (1)$$
$$L = \cup_\alpha L_\alpha.$$

Like ordinal definability, constructibility is nondiagonalizable in the following sense: if we form the constructible hierarchy and then add to the language of set theory the predicate "X is constructible," we do not obtain any new constructible sets. The property of being an ordinal is also absolute in the standard set-theoretical sense.

But Gödel does not see L as the right notion:

> . . . but, comparing constructibility with the concept of ordinal definability just outlined, you will find that not all logical means of definition are admitted in the definition of constructible sets. . . This has the consequence that you can actually define sets, and even sets of integers, for which you cannot prove that they are constructible (although this can of course be consistently assumed). For this reason, I think constructibility cannot be considered a satisfactory formulation of definability. (Gödel 1940)

Gödel seems to be claiming here, albeit in a roundabout fashion, that the Axiom of Constructibility is independent of the axioms of set theory – an extraordinary claim, if true. But priority of theorems is a minor concern. Gödel rejects L on the basis of its leaving out too many sets. We will see below that in some late remarks to Hao Wang, Gödel repeats this thought, drawing a distinction between what he calls "subjective mathematics," which

---

[39] According to Kreisel (1987, p. 158).

gives only a part of what we know, and "objective mathematics," which gives, so to speak, the rest. Gödel places constructibility on the "subjective" side.

Gödel ends the lecture with the remark that of his two candidates for the concept of absolute definability – constructibility and ordinal definability – neither is *truly* absolute:

> ... in both examples I gave, [ordinal definability and constructibility] the concepts arrived at or envisaged were not absolute in the strictest sense, but only with respect to a certain system of things, namely the sets as conceived in axiomatic set theory; i.e., although there exist proofs and definitions not falling under these concepts, these definitions and proofs give, or are to give, nothing new within the domain of sets and propositions expressible in terms of "set," "∈," and the logical constants. (Gödel 1940)

Gödel's point here is subtle: a concept of definability which is absolute "in the strictest sense" is necessarily fully transcendent, which is to say, formalism free. As in the case of the other epistemological notions, absolute definability in the strictest sense cannot be tied to definability in a logical calculus, be it the ZFC axioms of set theory or indeed any other theory or logical calculus.

As it happens, by an almost happy coincidence there is only one logical calculus on the horizon anyway, as far as Gödel was concerned: set theory, *always*.

2.5 Lawlikeness

We return to the idea of grounding the notion "comprehensibility by the mind." We hope that this idea has occurred to the reader: what is needed is a mathematical analysis as convincing as Turing's was, but of the notion of *lawlikeness*. Clearly what is comprehensible to us (to our minds) is what is regular, or lawlike; and indeed Gödel's analysis of definability comes close to offering us just that, a mathematical analysis of lawlikeness.

In terms of Gödel's philosophical influences, it is a solidly classical idea that nature is "lawful." For example, uniformity or purposiveness of nature is central to Kant's "Critique of Judgement,"[40] in which Kant distinguishes between various kinds of purposiveness, i.e., subjective and objective purposiveness, formal purposiveness and so on, according to whether such judgements are reflective or determinative, or teleological, or combinations of these. What is essential to each is the typically Kantian observation that

---

[40] From Ak. 180. Kant had a great influence on Gödel's philosophical outlook during the 1940s.

nature is *purposive for our cognition*; that is, its systematicity creates the transcendental ground for the possibility of cognition to be coherent. The purposiveness of nature, in Kant's terms, calls reason to operate as "a system of cognitions" rather than as a disordered "rhapsody of cognitions." This explains, for example, the possibility of science.

Kant explains purposiveness in subjective terms. As the true nature of things is necessarily walled away from us, so to speak, judgements of purposiveness cannot be judgements about actual, noumenal, states of affairs:

> Now this transcendental concept of a purposiveness of nature is neither a concept of nature nor a concept of freedom, since it attributes nothing whatsoever to the object (nature), but through this transcendental concept we only think of the one and only way in which we must proceed when reflecting on the objects of nature with the aim of having thoroughly coherent experience. (Ak. 185)

Limitations of space preclude a more extensive examination of lawlikeness. Our intention here was merely to point toward the way in which Gödel must have orientated himself philosophically, relative to the concept of lawlikeness.

2.6 *"Comprehensibility by the mind"*

In his *Logical Journey*, Hao Wang (1996) records a number of Gödel's very late remarks bearing on the issue of "comprehensibility by the mind."[41] In the interest of seeing how the issue of comprehensibility had ripened in Gödel's thinking by the end of his philosophical life,[42] we now examine some of these remarks, even so that limitations of space preclude our giving a very extensive analysis of them. The flavor of Gödel's remarks here, as well as the general framework, is very different from that of his published writings of the 1940s. Still, the programmatic element persists.

> To arrive at the totality of integers involves a jump. Overviewing it presupposes an (idealized) infinite intuition. In the second jump we consider not only the integers as given but also the process of selecting integers as given in intuition. 'Given in intuition' here means (an idealization of) concrete intuition...This is the beginning of analysis (of the concept of set). (Wang 1996, remark 7.1.11)

[41] The reader is referred to Wang's own qualifications on the faithfulness of these posthumously published transcriptions, see Wang (1996, p. xi).

[42] That is, by the early to mid-1970s.

Wang translates "idealized concrete intuition" as that intuition of finite sets which is not encumbered by any combinatorial complexity, and indeed for Gödel generally an idealization (of a particular kind) seems to be nothing more than the suppression of the restrictions associated with that particular case.

Relevant here is that idealizations for Gödel can be of two kinds, "objective" and "subjective," where idealized concrete intuition is categorized as "subjective." But the objective/subjective distinction is also one which Gödel applies to the mathematical field itself – wholesale, so to speak. Subjective mathematics is connected with surveyability – a kind of comprehensibility, surely – and objective mathematics is connected, of course, with truth:

> Subjectively a set is something we can overview in one thought. (Wang 1996, remark 8.3.2)

Gödel may use the term "overview" in the strict sense here, i.e., determining of each object individually that it belongs to the given set; or he may use "overview" in the Cantorian sense, i.e., grasping the set as a whole in one thought, or idea.[43]

In any case this act of subjective understanding, "the psychological act of thinking together all objects of a multitude" (Wang 1996, remark 8.3.3), "intuitive generation," in Gödel's terms, is a "basic principle of set formation" (Wang 1996, principle 8.7.1). What is relevant here is that the notion of set which comes closest to capturing this principle, which in fact bounds it, is constructibility:

> The proposition that all sets are constructible is a natural completion of subjective set theory for human beings. (Wang 1996, remark 8.3.5)

Thus for Gödel subjective set theory is that area of set theory generated by the surveyal of a unity in finite intuition, the idealization of which is expressed by the notion of constructibility. This gives a dividing line of sorts, between subjective and objective mathematics.

Gödel is clear in these remarks that the area of set theory based on subjective intuition is not sufficient to account for the whole picture; that is, subjective mathematics is not *mathematics*.

The undefinability of the universe V of all sets cannot be got by the subjective view at all.[44]

---

[43] In a slightly different context, in Wang (1996, remark 8.3.1), Gödel draws a distinction between these two, and uses the word "overview" in the first, strict sense.

[44] Wang (1996, remark 8.3.3). It is one of the many oddities of these transcriptions that Gödel does not mention Scott's result that constructibility is incompatible with large cardinals, see Scott (1961).

> To say that the universe of all sets is an unfinishable totality does not mean objective indeterminateness, but merely a subjective inability to finish it. (Wang 1996, remark 8.3.4)

> Even though for our knowledge we do bring in considerations of a more or less subjective nature, the range of possible knowledge is wider than the range of existence that can be justified from the subjective viewpoint... (Wang 1996, remark 8.3.3)

Indeed according to Gödel, subjective intuition does not even preclude the existence of inconsistent totalities:

> Without the objective picture, we do not seem able to exclude complete arbitrariness in determining when (the elements of) a multitude can be thought together (broader than can be overviewed) and when not... Indeed, without the objective picture, nothing seems to prevent us from believing that every multitude can be brought together... (Wang 1996, remark 8.3.1)

What of the objective picture? The undefinability of V, for example, can be proved rather easily, Gödel remarks. But such a proof involves the "objective":

> This significant property of certain multitudes—that they are unities—must come from some more solid foundation than the apparently trivial arbitrary phenomenon that we can overview the objects in each of these multitudes ... Some pluralities can be thought together as unities, some cannot. Hence there must be something objective in the forming of unities. (Wang 1996, remark 8.3.41)

The "objective picture" then, extends the "subjective picture"; it extends our concrete intuition, with reason acting normatively, although in a somewhat weak sense: reason gives, if not the *correct* extension, at least a consistent one (viz. computability). This is a familiar rationalistic idea in Gödel: reason singles out what is correct in our mathematical discourse.

How do we distinguish possibles from existents? What is needed to complete the "subjective picture" is to emphasize the "constitutive" aspect of our faculty of reason, and the related idea that the relation of mathematical objects to this faculty is itself objective and purposive. This is just what Gödel means, for example, in his remark that there is "something objective" in the *forming* of unities; the phenomenological idea that mathematical objects are constituted out of a purposive interplay between subjectivity and objectivity, with both the constituted object as well as the transcendental conditions for the object to be so constituted having objective properties:

3 Conclusion

We have given here a reading of Gödel (1946). A work rich in novel mathematical as well as philosophical insights, it was to be, for Gödel, among the last of its kind.[45] The methodological shift marked by the Gibbs Lecture, which we discussed above, was, for all intents and purposes, permanent. But so far the history of philosophy has not granted Gödel much in the way of recognition of his later philosophical work. In an attempt to cast a different light on this work, we hoped to point towords its programmatic origins.

---

[45] The so-called *Dialectica* interpretation was published in 1958, but the central ideas were based on work done in 1941.

I am very grateful to Menachem Magidor, whose conversations with me at the Mittag-Leffler Institute in the fall of 2009, and subsequently, were crucial for the development of the ideas of this chapter, as well as the related paper (Kennedy 2013), in connection with generalized versions of the constructible hierarchy and in the wider foundational context. Jouko Väänänen also influenced my thinking during this period, for which I am grateful. This interpretation of Gödel's 1946 Lecture was first presented at the 2009 Winter Meeting of the American Philosophical Association, as well as at numerous seminars and colloquia subsequently. I thank the audiences at those events for their stimulating questions and comments.

CHAPTER 7

Analyticity for realists

Charles Parsons

It is not easy to say what is the standing among philosophers today of the idea that mathematical truths are analytic, or what it has been since the decline of logical empiricism. After the debate touched off by W. V. Quine's "Two dogmas of empiricism" and "Carnap and logical truth" died down, the question of the analyticity of mathematics receded into the background. Some, like myself, accepted Quine's claim that the analytic-synthetic distinction could not do serious philosophical work in this domain. Others simply went in different directions.

In good part as a result of historical research, Carnap's philosophy has been viewed more sympathetically than it was during most of Quine's career after the publication of "Two dogmas." However, this has affected mainstream philosophy of mathematics only to a very limited degree. To the extent that there is a "received view" of the epistemology of mathematics, it is probably an empiricist view more or less on the model of Quine's. The radical pluralism implied by Carnap's position in *Logical Syntax of Language* has not found renewed acceptance. Historically, that was probably a product of the conflict between intuitionism, the views of the Hilbert school, and Russellian logicism that was so dominant in the inter-war period. Carnap's view of the relation of mathematics to empirical science has not dated in the same way, and it has had a strong influence on Michael Friedman's views on that subject. But I do not think he adopts a central part of Carnap's view connected with the analyticity thesis, the view that logic and mathematics are "tautological" or "without factual content."

A very preliminary version of this paper was presented to a conference at the University of Illinois at Chicago on September 24, 2011 honoring W. D. Hart. It was a great pleasure to honor Bill Hart, both for his contributions to philosophy and for his stimulating friendship over many years. Material from this paper was also incorporated into talks at Harvard University and the University of Oslo in November 2011. I am indebted to all these audiences. Thanks to Peter Koellner, Richard Tieszen, and the editor for comments on a more recent version.

I believe that the idea that some or all of mathematics is analytic has had some revival in more recent years, for example in the neo-Fregean program. However, since this chapter is about Gödel, my attention to this revival will be very limited. As we will see, Gödel had his own version of the claim that mathematical truths are analytic. After discussing Gödel's treatment and use of the notion and some problems it presents, I will argue that it has influenced the way some more recent writers have approached the question of the justification of the axioms of set theory.

1   Gödel's concepts of analyticity

In 1985, still influenced by the Carnap–Quine debate, Hao Wang undertook to introduce Kurt Gödel as a third party to that debate (Wang 1985). I had just begun to write about Gödel, although I would soon thereafter become an editor of his posthumous writings. Wang called attention to some underlying agreement between Carnap and Quine, although not everyone would agree with his characterization of it as "two commandments of analytic empiricism":

(a)   Empiricism is the whole of philosophy, and there can be nothing (fundamental) that can be called conceptual experience or conceptual intuition.
(b)   Logic is all-important for philosophy, but analyticity (even necessity) can only mean truth by convention (Wang 1985, p. 451).

Gödel's dissent from the first of these propositions is evident. His rather cryptic development of ideas about analyticity is certainly in a different direction from (b) or even from refinements of it to make it fit Carnap and Quine more accurately.

I do not propose to pursue Wang's thoughts on these matters further. But I will follow him in introducing Gödel as a third party to the discussion of analyticity in relation to mathematics. Gödel's earliest treatment of the matter is in some brief remarks near the end of "Russell's mathematical logic," where he asks whether and in what sense the axioms of *Principia* are analytic. He proposes two senses of this term and argues that they give opposite answers. The first is suggested by Leibniz's definition of necessary truth:

> First, it may have the purely formal sense that the terms occurring can be defined (either explicitly or by rules for eliminating them from sentences containing them) in such a way that the axioms and theorems become

special cases of the law of identity and disprovable propositions become negations of this law. (Gödel 1990, pp. 138–139)[1]

If arithmetic were analytic according to this criterion, then it would be decidable, and so Gödel concludes that it is not, and from the essential undecidability of arithmetic it follows that no stronger system can have its theorems analytic in this sense.

Gödel considers briefly but dismisses an infinitary version of this criterion, possibly suggested by Leibniz's treatment of contingent truth, in which in the process of reduction to identities one admits as intermediate steps sentences of infinite and even nondenumerable length. He says that the axioms of *Principia* are analytic by this version of the criterion, but then he says:

> But this observation is of doubtful value, because the whole of mathematics as applied to sentences of infinite length has to be presupposed in order to prove this analyticity, e.g., the axiom of choice can be proved to be analytic only if it is assumed to be true. (Gödel 1990, p. 139)

This remark seems to anticipate criticisms he made later of logical empiricist views of mathematics.[2]

Gödel then advances another, less clear criterion:

> In a second sense a proposition is called analytic if it holds "owing to the meaning of the concepts occurring in it," where this meaning may perhaps be indefinable (i.e. irreducible to anything more fundamental). (Gödel 1990, p. 139)

There seems to be a doubling in speaking of the "meaning of the concepts" in a proposition. However, better formulations occur in later writings. Thus in the Gibbs Lecture he says that a mathematical proposition "is true already according to the meaning of the terms in it, irrespective of the world of real things" (Gödel 1995, p. 320), and in Syntax V he says, "Mathematical propositions are true in virtue of the *concepts* occurring in them" (Gödel 1995, p. 357, emphasis Gödel's).

Before Quine's critique of analyticity, Gödel was almost alone in maintaining that the question whether mathematics is analytic might depend on one's choice among different senses of that term. In particular,

---

[1] Gödel's writings will be cited by volume and page number of the *Collected Works* (1986–2003). "Russell's mathematical logic" will be cited as RML; the Gibbs Lecture of 1951, "Some basic theorems of the foundations of mathematics and their philosophical implications" as Gibbs Lecture; "Is mathematics syntax of language?" of which six drafts exist, probably dating between 1953 and 1959, as Syntax followed by version number.

[2] See Parsons (1995b) reprinted in Parsons (2014).

this was not an option that Quine himself considered. Gödel seems to have thought that his first sense was at least as strong as Kantian analyticity, and he concluded that in Kant's sense arithmetic is indeed not analytic. Nowadays it would be a commonplace among Kant scholars that Kantian analyticity is quite narrow and that, looked at in that light, his arguments for the syntheticity of arithmetic cannot be dismissed. However, Gödel's argument does not give support to Kant's famous claim that "$7 + 5 = 12$" is synthetic, since the proposition belongs to a natural decidable fragment of arithmetic. He could well have agreed with Henri Poincaré's view that elementary identities of that kind are analytic while propositions requiring more than calculation to see their truth are synthetic.[3]

2 From the Russell paper through the 1950s

It is Gödel's second sense that is central to this paper, because only this sense offers the promise of rehabilitating the idea that either all of mathematics or at least some substantial part of it is analytic. From here on I will use such phrases as "Gödel's notion of analyticity" to refer to this sense. Already in RML Gödel says that maintaining the axioms of *Principia* to be analytic in this sense does not contradict the view expressed in most of the paper that "mathematics is based on axioms with a real content" (Gödel 1990, p. 139, note 47). The general idea is that if a statement is true by virtue of certain concepts, then the existence (and perhaps certain relations) of these concepts will itself be a substantial assumption (see Section 5 below.)

Gödel expresses his idea a little more fully in two papers he did not publish, the Gibbs Lecture of 1951 and "Is mathematics syntax of language?" He no longer talks specifically of the axioms of *Principia* but quite generally of "mathematical propositions."[4] Thus in the former place he writes that a mathematical proposition says nothing about the "physical or psychical reality existing in space and time" but is true by virtue of "the meaning of the terms in it" (Gödel 1995, p. 320). This meaning consists in certain concepts, which form "an objective reality of their own, which we cannot create or change, but only perceive and describe" (Gödel 1995, p. 320). A mathematical proposition "still may have a very sound objective

---

[3] Gödel seems to admit in the Gibbs Lecture that one could show that "$7 + 5 = 12$" is a tautology (Gödel 1995, p. 319). The context is a different argument from the one from undecidability considered in the text, and Gödel says that it would require a special definition of addition, restricted to a definite initial segment of the natural numbers.

[4] See also the passage from Syntax V (Gödel 1995, p. 357) quoted above.

content, insofar as it says something about the relations of concepts" (Gödel 1995, p. 320). The way this is manifested is that axioms of mathematics are not tautologies (or analytic in the narrower sense he had proposed) but still follow from the meaning of the primitive terms.

In this passage Gödel says straight out what is implied by the remark in RML, that his conceptual realism is what makes it possible to hold that mathematical propositions are analytic in the sense that concerns us. So it is not clear that a weaker form of realism, limited to mathematical objects or objectivity, would lead to that conclusion. However, already in this paper Gödel takes into account the possibility (mentioned elsewhere in RML but not in the remarks on analyticity) that mathematical axioms may be assumed "(like physical hypotheses) on the grounds of inductive arguments, e.g. their success in applications" (Gödel 1995, p. 347). Does that allow the possibility that mathematical conclusions that depend on axioms of that kind are *not* analytic, contrary to the very general claims Gödel makes in the Gibbs Lecture and the Syntax paper? I will not pursue that question now, but it will occupy us later.

In an earlier publication (Parsons 1995b) I pointed out some parallels in the arguments Gödel gives against the views of logic and mathematics he attributes to the Vienna Circle and arguments to be found in Quine's writings, especially "Carnap and logical truth." Supplementing Wang's remark on the agreement between Carnap and Quine, I think underlying these parallels is an important point of agreement of Quine and Gödel, the rejection of the view that mathematical statements are anything like tautologies and are void of substantive content. Gödel was concerned to argue that mathematics is based on axioms with a "real content."

In other respects Gödel's and Quine's philosophies could hardly be more different. The most striking difference is between Quine's naturalism and Gödel's anti-naturalism. One difference relevant to the present discussion is Gödel's realism about concepts. One cannot imagine Quine assenting to the following often quoted statement (in part already quoted) from the Gibbs Lecture:

> What is wrong, however, is that the meaning of the terms (that is, the concepts they denote) is asserted to be something man-made and consisting merely in semantical conventions. The truth, I believe, is that these concepts form an objective reality of their own, which we cannot create or change, but only perceive and describe. (Gödel 1995, p. 320)

Gödel's arguments concerning analyticity as they stand could only have force against the sort of critique that would be mounted by someone following Quine if he could give an adequate foundation for his reliance on this view of concepts. Neither Gödel's published writings nor his unpublished essays contain anything like a theory of meaning. In Parsons (1995b) I argued that he did not offer a foundation for the notion of concept that he was deploying. But he himself expressed dissatisfaction with his own thought about concepts and in 1959 turned to the study of Husserl.[5]

I do not have a clear picture of my own about what the study of Husserl did for Gödel's thought about concepts. Clearly, as Hauser (2006) argues, the idea of categorial intuition in Husserl's Sixth Logical Investigation offers a model for how Gödel could have come to understand perception of concepts.[6] But it would be on a general level and not say much about what conditions would be sufficient for the "perception" of such a concept as that of one or another large cardinal.[7]

3 Intuition and the view expressed in 1964

In later writings and in the conversations with Hao Wang, Gödel does not repeat the claim that mathematical truths are analytic in his second sense. Some of the slack is taken up by his notion of mathematical intuition, which first appears clearly in the Syntax drafts of the 1950s. It will turn out that there is a connection of the two notions. In the 1964 version of "What is Cantor's continuum problem?" analyticity is still present although more implicitly than in the texts canvassed so far. This is when he talks of what is "implied by" or "unfolds" the concept of set, as we shall see.

I have already written about Gödel's conception of mathematical intuition, and much of what I say here will merely summarize briefly what I have said elsewhere.[8] Gödel alludes to perception of concepts in one place

---

[5] On the dissatisfaction see the well-known letter to Paul Arthur Schilpp of February 3, 1959 (Gödel 2003b, p. 244).

[6] See Hauser (2006, §6). The relevance of this idea of Husserl's and the related discussion of "intuition of essences" in later writings had been explored in earlier papers by Richard Tieszen (1998, 2002), both reprinted in Tieszen (2005).

[7] In correspondence in February 2012, Dagfinn Føllesdal has pointed out that for Carnap, Quine, and for much of the discussion in the wake of Quine's critique, the analytic-synthetic distinction is a distinction of statements made in language. Quine's rejection of it is a result of his own theoretical reflection on language. Analyticity according to Gödel is a matter of concepts that do not depend on language. Føllesdal sees here an affinity between Gödel and Husserl. Time has not permitted me to explore this matter.

[8] Parsons (1995a), reprinted with postscript in Parsons (2014). The postscript concerns statements by others on the development of Gödel's platonism and is not relevant here. Sections 4–5 of the paper say something about the development of Gödel's conception, which will not be gone into here.

in RML (Gödel 1990, p. 151) and more conspicuously in the Gibbs Lecture. Those texts do not make clear how literally one is to take this choice of terms, but in later writings, already in the Syntax drafts, a more literal reading is appropriate. It is reasonable to suppose that that is what Gödel had in mind earlier, at least in the Gibbs Lecture.

The term "intuition" has been used both with noun phrase and with sentential complements, so that one could speak of intuition of objects and of intuition that something is the case, briefly intuition *of* and intuition *that*. Kant's *Anschauung* is historically a very important instance of intuition *of*, while the intuitions of a native speaker of a language about the grammaticality of sentences are a clear example of intuition *that*. If one's conception admits both, then they will be closely intertwined.[9] That appears to be the case with Gödel, but when the first of these grammars is deployed, Gödel usually uses the term "perception." In particular, that is the term he uses in relation to concepts.

Gödel uses the term "intuition" in cases of rational evidence, where the statement is not derived from others nor justified on a posteriori grounds, nor explained away by some such strategy as the conventionalism he opposes in the Syntax drafts. This is, I believe, the primary use of "intuition" in Gödel's writing. But he sometimes uses the term without attributing to it the epistemic force he would claim for it in the primary cases. There is no reason to think that on his view intuition is *ipso facto* knowledge, still less that we have an infallible mode of cognition by which we know mathematical axioms. He expects his opponents to concede the existence of intuition to the effect that certain statements are true or inferences valid, as in the following remark about intuitionism:

> The existence, as a psychological fact, of an intuition covering the axioms of classical mathematics can hardly be doubted, not even by adherents of the Brouwerian school, except that the latter will explain this psychological fact by the circumstance that we are all subject to the same kind of errors if we are not sufficiently careful in our thinking. (Syntax III, Gödel 1995, p. 338, note 12.)

A similar usage occurs in a well-known passage in the 1964 version of the continuum paper:

> The mere psychological fact of the existence of an intuition which is sufficiently clear to produce the axioms of set theory and an open series

---

[9] I attempt to sort out the philosophical uses of the term "intuition" in Parsons (2008, §§24–26).

of extensions of them suffices to give meaning to the question of the truth or falsity of propositions like Cantor's continuum hypothesis. (Gödel 1990, p. 268)

In these cases, intuition may convey a significant degree of plausibility (what I call intrinsic plausibility in Parsons 2008, pp. 319–320), but Gödel is evidently refraining from claiming that this amounts to being evident. For this reason, I do not think there should be dispute about the existence of what Gödel calls mathematical intuition, although he is acknowledging a dispute about its epistemic force, maybe especially in the case of higher set theory.

Gödel evidently thinks that in the salient cases of mathematical intuition or rational intuition in general, perception of concepts is involved. He claims this in a simple case that he uses to argue for one of his most contested claims, that there is a close analogy between mathematical or logical intuition and sense-perception:

> The similarity between mathematical intuition and a physical sense is very striking. It is arbitrary to consider "this is red" an immediate datum, but not so to consider the proposition expressing modus ponens or complete induction .... For the difference, insofar as it is relevant here, consists solely in the fact that in the first case a relationship between a concept and a particular object is perceived, while in the second case it is a relationship between concepts. (Syntax V, Gödel 1995, p. 359)

In version III of the same paper, he says that "with mathematical reason we perceive the most general (namely the 'formal') concepts and relations" (Gödel 1995, p. 354).

Gödel does not claim that perception of concepts is natural or immediate in the way that sense-perception is. It is in fact a consequence of his conceptual realism that in some cases clear perception of a concept is the result of considerable work, often over a long period. In fact, Wang records remarks that reject the model of conceptual clarification according to which one begins with a vague concept (such as that of computability) and through a process of clarification arrives at a sharp concept (computability by a Turing machine). Gödel's claim is rather that the sharp concept was there all along, but through the process of analysis came to be perceived more clearly.[10]

---

[10] See for example H. Wang (1974, pp. 84–85) and H. Wang (1996, remark 7.3.1). In remark 8.5.20 Gödel seems to reject the very idea of vague concepts: "Every concept is precisely defined, exactly and uniquely everywhere: true, false, or meaningless." However, Tieszen speculates that Gödel may have meant this remark to apply only to concepts of logic and mathematics.

Where most of us would speak of understanding of concepts, Gödel is prepared to use the language of perception. That results, however, from his insistence on a strong analogy between understanding of concepts and perception and between mathematical intuition and cognition by perception. In effect, he equates understanding of concepts with perception of them.

Gödel's emphasis on perception of concepts creates a link between mathematical intuition and analyticity. Mathematical intuition concerning a proposition rests on perception of the concepts in it. But that could only be sufficient for knowledge of the proposition if it is true by virtue of the concepts constituting it. But then it must be analytic in Gödel's sense. This reflection suggests the following conjecture:

> According to Gödel, a mathematical proposition is derivable from axioms knowable by intuition if and only if it is analytic.

A difficulty the conjecture faces is that analyticity is in the background in the 1964 Continuum paper, the writing of Gödel where intuition is most prominent. However, I will argue that it is present. It is also surprising that if the conjecture is correct, Gödel does not state it directly either in his known writings or in remarks reported by Wang. I will not try to decide here whether the conjecture is true, but it is plausible enough that it should be on the table for the reader's consideration.

Let us now turn to the 1964 version of the Continuum problem paper. Our observations should help us to understand the most quoted passage where Gödel invokes mathematical intuition, in the supplement to that paper:

> But, despite their remoteness from sense experience, we do have something like a perception also of the objects of set theory, as is seen from the fact that the axioms force themselves on us as being true. I don't see any reason why we should have less confidence in this kind of perception, i.e. in mathematical intuition, than in sense perception, which induces us to build up physical theories and to expect that future sense perceptions will agree with them, and, moreover, to believe that a question not decidable now has meaning and may be decided in the future. (Gödel 1990, p. 268)

By "the objects of set theory" Gödel means primarily concepts, the concepts of set and membership and probably the iterative conception of the universe of sets.[11] In an earlier footnote (Gödel 1990, p. 259, note 14) he speaks of an operation "set of x's," which he evidently views as a

---

[11] However, the "objects of set theory" presumably include sets. I do not believe that Gödel thinks we have "something like a perception" of the sets that are of concern to set theorists other than by way of concepts that define them. But firm textual evidence on his views about cognition of individual sets is lacking.

component of the iterative conception. In another footnote (Gödel 1990, p. 260, note 19), he mentions the concept "property of set," apparently making way for allowing higher-order logical operations in the language of set theory, since he mentions that it can be iterated.[12]

One can complain about the rather cryptic and dispersed manner in which Gödel describes the concepts that are central to set theory. It is also curious that Gödel lays great stress on the iterative conception of set, and the concepts that he mentions are, one might say, elements of that conception (at least as Gödel understands it); he does not list iteration itself among these elements. Martin (2005) has given a penetrating analysis of Gödel's use of the phrase "concept of set," pointing out that he very often means "concept of a (possible) universe of sets." This feature of Gödel's exposition is no doubt connected with the rather confusing presentation just commented on. It probably reveals some uncertainty on Gödel's part. I am not sure what it was; perhaps it was about what the real primitives of set theory should be.

However, this is apart from our main theme. There are two places that show the continued presence of his notion of analyticity.

First, Gödel observes that the "very concept of set" on which the axioms of set theory (by which he means ZF or ZFC) are based suggests "their extension by new axioms which assert the existence of still further iterations of the operation 'set of'" (Gödel 1990, p. 260). The examples he cites are inaccessible and Mahlo cardinals. Summarizing he says

> These axioms show clearly not only that the axiomatic system of set theory as used today is incomplete, but also that it can be supplemented without arbitrariness by new axioms which only unfold the content of the concept of set explained above. (Gödel 1990, pp. 260–261)[13]

By the latter concept he means the iterative conception of set or, better, of the universe of sets. In a footnote Gödel mentions stronger axioms of infinity that had been discussed very recently, in particular the existence of a measurable cardinal, and remarks, "That these axioms are implied by the general concept of set in the same sense as Mahlo's has not been made clear yet" (Gödel 1990, p. 260, note 20).

Second, he speaks of the "intrinsic necessity" of an axiom, and it is plausible that he means its being implied by the general concept of set. But

---

[12] However, he cautiously remarks in the same note, "The new axioms thus obtained, however, as to their consequences for limited domains of sets (such as the continuum hypothesis) are contained (as far as they are known today) in the axioms about sets."

[13] Remarks to the same effect occur in the 1947 version (Gödel 1990, pp. 181–182).

at this point Gödel mentions the possibility of axioms that may not be intrinsically necessary but may have an inductive justification by virtue of fruitfulness of their consequences. In a nutshell,

> There might exist axioms so abundant in their verifiable consequences, shedding so much light upon a whole field, and yielding such powerful methods for solving problems ... that, no matter whether or not they are intrinsically necessary, they would have to be accepted at least in the same sense as any well-established physical theory. (Gödel 1990, p. 261)

This suggestion was not new; it was already present in RML (as noted above), and even the formulation is mostly taken over from the 1947 version (Gödel 1990, pp. 182–183). So it was not Gödel's view that acceptance of an axiom required that it be "implied by the concept of set" and thus analytic. But that it "had not been made clear yet" that what are now called large cardinal axioms are implied by the concept of set indicates that he still at least hoped that this would be made clear. He may even have thought that they *were* "implied by the concept of set," but that we did not yet see how that is so. But the remark just quoted indicates that he did not regard this as essential for the acceptability of an axiom.

This does represent some backing away from the view apparently embraced in the Gibbs Lecture and in Syntax, that all mathematical truths are analytic. It is possible that he had come to regard it as an ideal that may or may not be realized. Gödel hints that what has come to be called extrinsic justification does not give quite as strong a reason for accepting an axiom as its being seen to be implied by the concept of set. Thus studying the "success" of an axiom can yield "a probable decision concerning its truth," and in the hypothetical case described in the quotation, the axioms "would have to be accepted at least in the same sense as any well-established physical theory."

4 After 1964

Indications about Gödel's later views on the question of the analyticity of statements in higher set theory are sparse and not very conclusive. The only source published in Gödel's lifetime is those remarks in Hao Wang (1974) *From Mathematics to Philosophy* that are attributed to him. Wang says explicitly that they were approved by Gödel, and other evidence indicates that he was responsible for their final revision.[14] The first of the

---

[14] H. Wang (1974, p. x). On the remarks on set theory (pp. 186 and 189–190), see Gödel's letter to Ted Honderich of June 27, 1972 (Gödel 2003b, pp. 77–78); on Gödel's role in the final revision of the book see §2.3 of my introductory note to the Gödel–Wang correspondence (Gödel 2003b, pp. 388–389).

relevant remarks on set theory, on the justification of the axiom of replacement, certainly represents the axiom as implied by the concept of set, but clearly it tells nothing about his view of large large cardinal axioms.

The five principles for setting up axioms of set theory (pp. 189–190) are not explicit on this point, but it is hard to believe that principles (2)–(4) are not implications of the iterative conception as Gödel understood it. Principle (1), "Existence of sets representing intuitive ranges of variability," is one that I have had great difficulties with,[15] but there is some plausibility in supposing that Gödel thought that it too is such an implication. But the case of Principle (5), the uniformity of the universe of sets, does not strike me as at all clear. If Gödel thought the iterative conception implied that a property occurring at one stage of the iterative hierarchy must recur at higher stages, possibly in a modified form, one would expect an argument. I do not know of one in Gödel's writings.

For further illumination of Gödel's late views we must look to Hao Wang's reports of their conversations. Analyticity is not mentioned by name in the remarks reported by Wang. Although it is still implicitly present, it is less so than in 1964. Perhaps the most relevant passage is remark 8.7.16, which comes at the end of a discussion of the set theories of Ackermann, Powell, and Reinhardt. Gödel states what he says is a "reasonable formulation" and then says:

> Generally, I believe that, in the last analysis, every axiom of infinity should be derived from the (extremely plausible) principle that V is indefinable, where definability is to be taken in [a] more and more generalized and idealized sense. (Wang 1996, p. 285)

The suggestion is that axioms of infinity "should be" obtained by an appropriate generalization of reflection principles. Although the context of this remark is Reinhardt's development of Ackermann's ideas to obtain very large cardinals, Reinhardt himself reports that Gödel did not think his motivation of the axioms he proposed adequate (Reinhardt 1974, p. 189, note 1).

Earlier in this chapter Wang makes some remarks, not claimed to be reports of specific statements by Gödel, offering an interpretation of Gödel's view on these issues:

> In his discussions with me, Gödel stressed the central importance of the axiomatic method for philosophy. He did not elaborate his conception of

---

[15] See Parsons (1977), at pp. 275–280 of the reprint in Parsons (1983). In response, Wang clarified his position in "Large sets" (Wang 1977). I comment further on the matter in Parsons (1998). This paper is also reprinted in Parsons (2014).

the method except that he often gave the impression that the task is to find the primitive concepts and then try to see the true axioms for them directly by our intuition. In practice, of course, he recognizes that considerations on many levels are involved when we try to find the axioms or the principles of an area. Nevertheless, he seems to have, or assume, a notion of intrinsic necessity as the attainable ideal. (Wang 1996, p. 244)

It is clear that Wang is not only referring to the case of set theory, but he means the remark to apply to that case. The very next paragraph is devoted to quoting from the well-known remarks about extrinsic justification from the 1964 Continuum paper (Gödel 1990, p. 261).

The amount in Wang's report that bears on these issues at all is small; it is quite possible that the issue was not at the center of Wang's own attention, particularly in the later conversations after the publication of *From Mathematics to Philosophy* (H. Wang 1974).

5 Problems concerning analyticity and justification

It was noted early in the above discussion that Gödel thought that holding that mathematical propositions are analytic is compatible with the realism that he espoused. It is also clear that he saw realism about concepts as the key to this. In particular, he thought that the latter view included the claim that concepts are objective not just in the sense of being common to different thinkers when they understand each other but that they are not in any way human creations.

A point we have passed over is exemplified by one of Gödel's earliest remarks about the concept of analyticity that concerns us. In the course of arguing in RML that the axioms of *Principia* are, on a certain interpretation,[16] analytic, he says:

> This view does not contradict the opinion defended above that mathematics is based on axioms with a real content, because the very existence of the concept of e.g. "class" constitutes already such an axiom; since, if one defined e.g. "class" and "∈" to be "the concepts satisfying the axioms," one would be unable to prove their existence. (Gödel 1990, p. 139, note 47)

Gödel opts here for a stronger interpretation of the "existence" of the concept than that phrase might suggest. It is easier to see this in the simpler case of second-order arithmetic. The concept in question would, on the face of

[16] Gödel excludes the axiom of infinity and interprets "predicative function" as "class," so that the theory he has in mind is effectively the simple theory of types without the axiom of infinity. The note quoted in the text seems to gloss over the stratification into types; probably the metalanguage in which the "axiom" of the note is stated is to contain transfinite types.

it, be that of a structure consisting of a set N, an initial element a, and a function f playing the role of successor, such that f maps N one-one into itself with a not in its range, and the second-order induction axiom holds. On Gödel's view that predicates designate concepts, it seems that our ability to talk meaningfully of such a structure implies the existence of the concept at issue, essentially Dedekind's concept of a simply infinite system. But that leaves open the possibility that it might be vacuous: although the *concept* exists, it has no instances. But it is clear that that is not what Gödel means. The issue about existence in the remark about *Principia* quoted above is about "the concept satisfying the axioms," and I think Gödel means to say that if there is such a concept, then the axioms are not vacuous.

A related passage occurs in the Gibbs Lecture, where his focus is on the concept of set of integers:

> For example, the basic axiom, or rather, axiom schema, for the concept of a set of integers says that, given a well-defined property of integers (that is, a propositional expression $\varphi(n)$ with an integer variable n) there exists a set M of those integers which have the property φ. (Gödel 1995, p. 321)

In this passage Gödel is most concerned to insist that these axioms are not tautological, but he does say that they are "valid owing to the meaning of the term 'set'—one might say they express the very meaning of the term 'set'—and therefore they might fittingly be called analytic" (Gödel 1995, p. 321). In this case Gödel does not talk of the existence of a concept, where a single concept satisfying all these axioms would apparently have to involve the concept of truth, which Gödel does not mention. It may be that he thought that the talk of the "existence" of a concept could lead to misunderstanding.

In the *Principia* case, the existence assertion would have to be of a concept of higher type than the theory itself uses; it would apparently have to belong to ωth order logic. Perhaps one can see things most clearly in the case of a first-order theory. On Gödel's view, the predicates designate concepts that apply to objects in the domain. One may assume an additional concept for the domain. However, we can conceive a type 2 (or second-level) concept that applies to these concepts if and only if they satisfy the axioms. That the first-level concepts exist seems to follow just from Gödel's assumptions about meaning. I think that if the second-level concept is described in the way I have just done, that also follows. But what corresponds to the existence assertion about *Principia* in RML is the statement that there are concepts that *do* satisfy the axioms.

Martin questions whether such statements are analytic, and on their face they do not satisfy the condition of being "implied by" the concepts of the theory. The problem recalls the well-known difficulty about the onto-logical argument for the existence of God: how can a concept imply that there exists an instance of it? One might reply that since these existence statements contain additional concepts, they might be true by virtue of the concepts constituting *them*. But it is not evident that this is the case.

In the remarks from 1964 that we discussed in Section 3, existence of relevant concepts is not explicitly mentioned. It is most instructive to follow Martin and assume that the "concept of set" at issue is generally that of a structure answering to the iterative conception. If it is implied by that concept that such a structure contains inaccessible and Mahlo cardinals, then it seems that there is still a question close to the one that Gödel raised in the earlier context about existence. And indeed, Martin regards it as a serious question whether the "concept of set" so understood is instantiated (Martin 2005, pp. 220–221). Although this goes beyond the existence of the concept in the weaker, more straightforward sense, it fits with Gödel's stronger understanding in the passage from RML.

In the cases we considered earlier, the concept whose "existence" is at issue is one that satisfies explicitly stated conditions, and what is true about the relevant structure is what follows logically from these conditions (though possibly in higher-order logic). The universe of sets is much more open ended than that, because the hierarchy of ranks is "absolutely infin-ite." Gödel's own example of inaccessible and Mahlo cardinals is one where the concept "implies" the existence of ranks well beyond what can be derived from the standard axioms. As we have seen, Gödel's views on how far what is implied by the concept reaches are not clear, and the matter has been a subject of some dispute since.

Martin maintains that the concept is sharp enough so that it can be shown to be categorical. It is probable that Gödel would have agreed with this. But it is just that that raises the question for Martin of whether the concept is instantiated, because categoricity implies that the continuum hypothesis has a determinate truth-value.[17] Martin then suggests that what matters for mathematics (at least pure mathematics) is what can be drawn out as implications of a concept, and for this it is not necessary that the concept should be instantiated, although he evidently thinks its consist-ency is necessary.[18] One would have the paradoxical consequence that set

---

[17] Martin does not deny that the continuum hypothesis has a determinate truth value. In correspondence in August 2010, he stated that he was open minded about the question.
[18] He also mentions "coherence or 'soundness'" (Martin 2005, p. 220) but does not say explicitly that they are requirements for the mathematical usefulness of a concept.

theory can be an entirely legitimate enterprise, even though in a sense there are no sets. This was surely not Gödel's view. Although it is not clear how far Martin embraces Gödel's realism about concepts, for Gödel himself realism also encompasses realism about objects. And if there is a distinction between the existence of the concept (in the weaker sense) and its being instantiated, how can mathematical intuition, which at bottom is perception of concepts, yield knowledge of mathematical truths? But I do not claim that Martin thought that this aspect of his view would have met Gödel's assent.

6 The notion of intrinsic justification

At the beginning of this chapter, I recalled the controversy about the analyticity of mathematics and said that it had effectively died out some years ago. However, in more recent years writers on set theory have used the term "intrinsic justification" in a way that seems consciously to echo the remarks of Gödel about the intrinsic necessity, or lack of it, of axioms. We have maintained above that in Gödel's usage intrinsic necessity amounts to analyticity in his second, broader sense.

It should be said that in recent usage the terms "intrinsic necessity" and "intrinsic justification" do not have a very fixed meaning. Thus commenting on Zermelo's remarks about the axioms he introduced in 1908, Penelope Maddy writes:

> ... this general defence is not phrased in terms of self-evidence or intuition or conforming to some underlying concept of set or anything else along such lines. Instead, the axioms are evaluated in terms of their consequences, or more broadly, in terms of the theory they produce. In what follows, I draw a rough-and-ready distinction between justifications of the former variety—intrinsic justifications—and those of the latter variety—extrinsic justifications. (Maddy 1997, p. 37)[19]

In her more recent book *Defending the Axioms*,[20] Maddy (2011) presents on p. 47 a rather cryptic characterization of the distinction that is probably meant to be equivalent to the earlier one.

[19] Even with this rather inclusive characterization of intrinsic justification, the survey of justifications of particular axioms in the literature that Maddy then undertakes reveals a surprising amount of extrinsic argument, not all of it, however, fitting the paradigm of extrinsic justification suggested by the remarks of Gödel quoted above.
[20] She ends this work with an argument for the claim that extrinsic justifications are more fundamental. I am inclined to think that that has always been her view.

Other writers seem to follow Gödel more closely in maintaining that an axiom is intrinsically justified only if it is "implied by the concept of set." The point of interest to us, however, is not to explore whether there are other kinds of justification that might reasonably be called intrinsic, but to explore intrinsic justification on this narrower understanding, to see whether the problems raised by Gödel's conception of analyticity are of current relevance.

I will begin this discussion with a simpler case, that of the natural numbers, and motivations offered for the axiom or axiom schema of induction. The intuition of many is surely that this principle is implicit in the concept of natural number or holds by virtue of the meaning of "natural number." Of course there is an easy way of cashing in this intuition, by giving a definition in second-order or set-theoretic terms from which the induction principle follows either by second-order logic or by elementary set theory. But the intuition has force also for those, among whom I number myself, who would prefer not to introduce second-order logic or set theory in setting forth the elements of number theory, and who also would not wish to be committed to Gödel's conceptual realism.

My own way of dealing with this issue was to begin with an explanation of the notion of natural number (that is to say of the predicate "natural number") in which induction is directly embodied,[21] to be sure with the gloss that it is a way of capturing the idea that the numbers are what is obtained by beginning with 0 and iterating the successor operation.[22] Induction is, thus, given by the explanation. But that does not rule out the possibility that quite different explanations would give rise to induction in a different way. In addition to the approach by second-order logic, there is the category-theoretic explanation that takes iteration of a function as fundamental.[23]

In what sense do considerations like these offer a justification of mathematical induction? There is certainly an issue corresponding to the one Gödel raises about the existence of relevant concepts. The explanations do not close off questions of whether we really understand the predicate explained in one of these ways, or guarantee that there will be consistency in the results of applying it. Tait points out that there is scope for disagreement about the application of the terms used in such explanations,[24] for example

---

[21] As an open-ended generalization over predicates, i.e., linguistic items.
[22] See Parsons (2008, §47), also Parsons (1992) especially pp. 141–143, 155–156.
[23] For a brief explanation see Tait (1986), p. 85, note 9 of the reprint in Tait (2005). The proposal evidently originated with F. W. Lawvere (1964).
[24] Tait (2005, p. 85), where the reference is to Dedekind's explanation or to constructivism. In the latter case the Lawvere/Tait explanation is available.

in the one I set forth, what is to count as a "well-defined predicate." In particular, the explanation does not by itself give assurance that any system of arithmetic is consistent. Moreover, the open-endedness implies that well-defined predicates can incorporate other notions, for example from set theory.[25]

However, such explanations surely do have some force in making clearer why mathematical induction should be evident. Moreover, they pose a challenge to someone who is skeptical about induction. One can ask him, "How do *you* understand the notion of natural number?" If he has an answer at all, it may offer the proponent a chance to argue that he too is committed to induction.

I am not sure how widely the term "intrinsic justification" is used in the narrower sense with respect to axioms of set theory. Two writers who do seem to use it in this sense are William Tait and Peter Koellner. Tait has in several publications undertaken to obtain cardinal numbers in a way that directly develops the iterative conception of set as he understands it. However, he wants to avoid one feature of Gödel's view, because he rejects the idea that there is a single determinate total universe of sets.[26] He also undertakes to avoid a feature common to many standard expositions of the iterative conception: assuming at the outset "stages" at which sets are "formed." The stages are in effect ordinals, so that it at least appears that ordinals are being assumed outside the process of generation of sets. There are two items that he does take as belonging to the conception: second-order logic, and Cantor's principle that a set of ordinals has a least upper bound. He says he is assuming at the outset the "more primitive, logical notion of set" embodied in second-order logic (Tait 2005, p. 134). That is the notion of arbitrary subset of a given domain of objects.[27] At the time of these writings he maintained that extrinsic justification of axioms contradicted the iterative conception. Commenting on Gödel's well-known remarks of 1964 about extrinsic justification, he writes:

> This criterion seems to have had some influence among logicians studying modern set theory. But it is difficult to reconcile it with the iterative

---

[25] Peter Koellner (2009, p. 207 and note 5) gives the consistency of ZF plus the axiom of determinacy as an example of a statement in the language of arithmetic that is presumably true but does not have an intrinsic justification. He says that it is not intrinsically justified on the basis of the conception of natural number. I doubt that he would say that it is intrinsically justified on the basis of the conception of the universe of sets.

[26] See "Constructing cardinals from below," in Tait (2005, pp. 133–154, at p. 142) or Tait (1998, p. 282).

[27] This is, I believe, close to what Gödel, in his remarks on the iterative conception, describes as the operation "set of" or "set of *x*'s."

conception. On the latter conception, the "instrinsic necessity" of an axiom arises from the fact that it expresses closure under some operation that we have obtained for constructing domains or ordinals. To introduce a new axiom as "true" on this conception because of its "success" would have no more justification than introducing in the study of Euclidean space points and lines at infinity because of their success. One may obtain an interesting theory in this way and one worthy of study, but it will not be Euclidean geometry. A "probable decision" can only be a probable decision about its derivability from that conception. (Tait 1998, p. 275)[28]

The view expressed in this passage is not Tait's present view. The point of quoting it is to document the claim that at the time he wrote it Tait understood the justification of axioms of set theory to be intrinsic justification in the narrower sense that is our focus. Other writings of Tait indicate that he does not commit himself to Gödel's conceptual realism, at least in its full-blown form. But these remarks and the analysis of set-theoretic concepts that lies behind them at least raise a question as to whether he was committed to a notion close to Gödel's notion of analyticity.

This suggestion should be qualified in light of Tait's general view of the role of axioms in mathematics. Tait defends what he calls the axiomatic conception of mathematics, according to which the criterion of truth in mathematics is proof from axioms, but we should not ask of axioms that they should have some epistemologically fundamental property that theories of rational or intuitive evidence have sought to capture.[29] A variety of considerations can be offered in favor of axioms; they belong to what Tait calls "dialectic," in a sense he derives from Plato. It follows that the considerations from his interpretation of the iterative conception of set belong to dialectic. Arguments to the effect that certain axioms are implicit in that conception may offer good reasons for accepting them, but once the axioms are accepted the only grounds on which a statement can be accepted as true is proof from the axioms. In setting forth this conception, Tait seems to allow for a variety of considerations in favor of axioms and to reject the idea that an axiom needs to be evident in any intrinsic way.

In two publications, Peter Koellner derives his use of the term "intrinsic justification" from Gödel and thus has in view the narrower sense that concerns us.[30] However, it is not clear how much he means to commit himself to in using this notion. In the earlier paper, the aim is mainly to explicate and reconstruct some views of Gödel. And in both papers he seems

---

[28] Cf. Tait 2005, pp. 283–284.   [29] See especially the introduction and Essay 4 of Tait (2005).
[30] Koellner (2006), reprinted with revisions in Feferman *et al.* (2010), and Koellner (2009).

to assume for the sake of the argument that certain justifications are intrinsic and then to argue for a thesis about the limits of intrinsic justification, at least as far as that notion is understood today. He is very cautious about any positive claims of intrinsic justification. Thus he says that the iterative conception of set "(arguably) justifies instances of Replacement and Comprehension for certain extensions of the language of set theory" (Koellner 2009, p. 207). About the reflection principles that are his main focus, he says that they are "the best current candidates for axioms that admit such an intrinsic justification" (Koellner 2009, p. 208). It is very likely his concern with the limits that leads him to be generous about what counts as a candidate and admit at the outset all the higher-order reflection principles introduced by Tait. But he raises some philosophical difficulties about this proposal and concludes that "at the moment we do not have a strong intrinsic justification of higher-order reflection principles" (Koellner 2009, p. 209). The sharper conclusion that he offers is that some of the principles are implied by the existence of the Erdös cardinal $\kappa(\omega)$ and are thus consistent with $V = L$ (and so cannot imply the existence of "large" large cardinals or that of $0^{\#}$), while stronger ones are inconsistent.

Koellner (2009) supplements his case by arguing in §7 that the principles introduced by Reinhardt differ in essential ways from more straightforward reflection principles and there is less reason to take them to be intrinsically justified.[31]

These examples show that appeals to intuitions about what is implicit in a concept or conception are still made. In none is it claimed that spelling out such intuitions gives a complete justification of the axioms concerned. Issues connected with Gödel's concern about the "existence" of the concepts involved do not go away. Koellner regards intrinsic justification as only "intended to be definitive modulo the tenability of the conception" (2009, p. 207). Tait has the same view, but in his discussion of set theory it is more implicit.

Thus there is an unavoidable element of the extrinsic in justifications of this general kind. Gödel seems not to have admitted this explicitly, and his conceptions of intuition and perception of concepts might be a strategy for avoiding it. But he made very clear that one can perceive concepts more or less clearly, and experience with reasoning with given concepts, in other words the development of theories, is a major way of arriving at such clear perception. So such an admission may be unavoidable even for Gödel.

---

[31] Reinhardt's ideas are set forth in Reinhardt (1974).

PART IV

The set-theoretic multiverse

CHAPTER 8

Gödel's program

John R. Steel

1 Introduction

In the nineteenth and early twentieth centuries, it was shown that all mathematical language of the time could be translated into the language of set theory (LST), and all mathematical theorems of the time could be proved in ZFC. A century later, mathematicians have yet to develop any mathematics that cannot be expressed in LST, and there are probably few who believe that this will happen any time soon. However, the remarkable work of Kurt Gödel in the 1930s, and Paul Cohen and his successors in the period from 1963 to the present, has shown that ZFC is incomplete in significant ways. There are very concrete statements about natural numbers that it fails to decide. There are conceptually central questions about real numbers and sets of real numbers that it fails to decide. Although mathematicians can say everything they have to say in LST, they cannot decide all the questions they would like to decide using only the axioms of ZFC.

Gödel anticipated that Cohen's theorem on the independence of the continuum hypothesis would eventually be proved. In his 1947 paper "What is Cantor's continuum problem" (Gödel 1947), he suggested a program of research now known as *Gödel's program*:

> Decide mathematically interesting questions independent of ZFC in well-justified extensions of ZFC.

The mathematical question Gödel had foremost in mind was the continuum hypothesis (CH), and the well-justified extensions of ZFC he had in mind were those that begin with ZFC + Con(ZFC), and make their way up through the hierarchy of strong axioms of infinity.[1]

---

[1] Gödel did not formulate his program in the broad terms we have used. But Gödel (1947) is the earliest paper we know of advocating this response to the incompleteness of ZFC with respect to the CH, and the program as we have stated it is an obvious extrapolation. See also Gödel (1946).

Gödel's program is inevitable, if you believe, as Gödel did, that the axioms of ZFC describe a "well-determined reality," or in less colorful language, that the meaning we have assigned to sentences in the syntax of LST suffices to determine a truth value for CH.[2] The qualification *well-justified* in the statement of Gödel's program also introduces a philosophical component. How does one justify statements in LST? General philosophical questions concerning the nature of meaning, evidence, and belief rear their ugly heads. Of course, the same questions come up vis-à-vis whatever fragment of LST and ZFC one might choose as a safe retreat. Moreover, for those who accept some reasonable fragment of ZFC, but believe that the truth value of CH is not determined by the meaning we currently assign to the syntax of LST, the continuum problem does not disappear. Certainly we do not want to employ a syntax which encourages us to ask pseudo-questions, and the problem then becomes how to flesh out the current meaning, or trim back the current syntax, so that we can stop asking pseudo-questions.

For these reasons, work in Gödel's program provides an interesting case study for the epistemology of mathematics, and perhaps epistemology will be able to give something back, for it seems likely that any solution to the continuum problem must be accompanied by a better understanding of what it is to be a solution to the continuum problem.

In the author's opinion, the key methodological maxim that epistemology can contribute to the search for a stronger foundation for mathematics is: *maximize interpretative power*. As Galileo put it in his well-known dictum, mathematics is the language of science. Our foundational language and theory should enable us to say as much as possible, as efficiently as possible. A set theorist cannot help but be aware that there are limits to the usefulness of such maxims, and that proofs usually count for more than plausibility arguments.[3] But the idea that set theorists ought to seek a language and theory[4] that maximizes interpretative power seems to carry us a long way. We shall discuss just how far it goes in this chapter.[5]

---

[2] I am using LST to stand for the usual syntax, together with the meaning we currently assign to it. This is why it makes sense to talk about translating pre-set-theoretic language into LST. What the translation preserved was meaning; meaning is that which is preserved by a good translation.

[3] In Einstein's words, "A scientist ... must appear to the systematic epistemologist as a type of unscrupulous opportunist."

[4] We believe the two evolve together.

[5] *Maximize interpretative power* has a lot in common with the old idea that in mathematics, consistency guarantees existence. The trouble with the old formulation is that consistency is a property of theories, not objects. If we try to use the old formulation to guide us to a foundational theory, we face the seemingly hopeless task of making useful sense of "objects that *T* is about."

2 The consistency strength hierarchy

One thing set theorists have understood much better in the years since Gödel (1947) is the family of possible extensions of ZFC. At one level, this family is rich indeed. There is a plethora of seemingly natural, basic questions about uncountable sets which ZFC does not decide. But underlying this great variety of consistent extensions of ZFC, and the corresponding wealth of models of ZFC, there is a good deal more order than might at first be apparent. To explain this, we must introduce the consistency strength hierarchy.

Definition 2.1 *Let T and U be axiomatized theories extending* ZFC; *then* $T \leq_{Con} U$ *iff* ZFC *proves* $Con(U) \Rightarrow Con(T)$. *If* $T \leq_{Con} U$ *and* $U \leq_{Con} T$, *then we write* $T \equiv_{Con} U$, *and say that* T *and* U *have the same consistency strength, or are* equiconsistent.

There is an intensional aspect here, in that the order really is on presentations of theories, rather than theories, but we shall ignore that detail here.

Of course, it was Gödel who discovered that there is no largest consistency strength. One can increase the consistency strength of T by adding to it $Con(T)$, or equivalently, there is a model of T. One can take larger steps upward by requiring that this model be arithmetically correct, or well founded, or an initial segment of the cumulative hierarchy. Still stronger reflection principles lead into the large cardinal hierarchy.

A terminological digression: some people would rather use "reflection principle" in such a way that large cardinal hypotheses at the level of measurables and beyond do not qualify (cf. Koellner 2009). I prefer the more liberal usage. It is the hierarchy of principles that starts with Con(ZFC), or Con(PRA) for that matter, and goes on through the existence of rank-to-rank embeddings that is the important natural kind. *Strong axioms of infinity* is a traditional term, and perhaps the best, but it is a mouthful. I shall tend to use *large cardinal hypothesis* below, although I mean to include here some statements that do not assert the existence of a cardinal number, such as "$O^{\#}$ exists."

Large cardinal hypotheses play a very special role in our understanding of the consistency of theories extending ZFC. Many natural extensions T of ZFC have been shown to be consistent relative to some large cardinal hypothesis H, via the method of forcing. This method is so powerful that, at the moment, we know of no interesting T extending ZFC which seems unlikely to be provably consistent relative to some large cardinal hypothesis via forcing. Thus the extensions of ZFC via large cardinal hypotheses seem to be cofinal in the part of the consistency strength order on extensions of

ZFC which we know about. These days, the way a set theorist convinces people that T is consistent is to show by forcing that $T \leq_{\text{Con}} H$ for some large cardinal hypothesis H. He probably had heuristic reasons to believe in advance that T is consistent, but this was the proof. Of course, such a proof only carries weight if there is evidence that H itself is consistent.

We do have pretty good evidence that even quite strong large cardinal hypotheses like the existence of rank-to-rank embeddings are consistent with ZFC. For supercompacts, the evidence is stronger, and when we move down to large cardinal hypotheses like the existence of Woodin cardinals, for which we have an inner model theory, it is much stronger still. In all cases, the evidence is basically the existence of a coherent theory in which the hypothesis plays a central role, a theory that extends in a natural way the theory we obtain from weaker hypotheses.[6]

The evidence from inner model theory is especially strong. If H has an inner model theory, then by assuming H we can give a systematic, detailed description of what a minimal model satisfying H might look like. The prototypical such model is Gödel's universe L of constructible sets, which satisfies ZFC, and various weak large cardinal hypotheses. Larger canonical models satisfy hypotheses like "there is a measurable cardinal" and "there is a Woodin cardinal." Canonical inner models admit a systematic, detailed, "fine structure theory" much like Jensen's theory of L. Such a thorough and detailed description of what a universe satisfying H might look like provides evidence that H is indeed consistent, for a voluble witness with an inconsistent story is more likely to contradict himself than a reticent one. Moreover, inner model theory lets us prove consistency strength lower bounds $\text{Con}(T) \Rightarrow \text{Con}(H)$ for various T which we may have some independent reason to believe consistent, and this provides further evidence that H is consistent. For example, Δ^1_2-determinacy, the existence of a saturated ideal on ω_1, and the existence of one Woodin cardinal are all equiconsistent over ZFC. The consistency of all of them follows from the consistency of the Proper Forcing Axiom. None of these statements has much to do with any of the others; there are different intuitions, motivations, and connections that led set theorists to them. This state of affairs makes it more likely that ZFC + "there is a Woodin cardinal" is consistent.[7]

---

[6] With regard to rank-to-rank embeddings, I am thinking of Woodin's theory of $L(V_{\lambda+1})$. With regard to supercompact cardinals, there are the many uses of them in forcing, by many people over a period of decades.

[7] We do not as yet have a full inner model theory for large cardinal hypotheses at the level of superstrong or supercompact cardinals. The author believes that one day we will.

Often, a consistency strength upper bound for T obtained by forcing can, with additional work, be made optimal. That is, we can find a large cardinal hypothesis H such that $T \equiv_{\mathrm{Con}} H$. The proof that $H \leq_{\mathrm{Con}} T$ involves constructing inside any model of T a canonical inner model of H. Unfortunately, although the basic theory of forcing is well understood, inner model theory lags well behind. At the moment, it cannot produce any nontrivial consistency strength lower bounds of the form $H \leq_{\mathrm{Con}} T$, where H is significantly stronger than "there is a Woodin limit of Woodin cardinals," a hypothesis of only middling strength.

The large cardinal hypotheses are themselves well ordered by consistency strength.[8] The pattern described in the last two paragraphs then leads to the following vague conjecture.

Natural consistency strengths are well ordered If T is a natural extension of ZFC, then there is an extension H axiomatized by large cardinal hypotheses such that $T \equiv_{\mathrm{Con}} H$. Moreover, $\leq_{\mathrm{Con}}$ is a pre-well order of the natural extensions of ZFC. In particular, if T and U are natural extensions of ZFC, then either $T \leq_{\mathrm{Con}} U$ or $U \leq_{\mathrm{Con}} T$.

It is difficult to see how one could make the conjecture more precise. One can construct unnatural extensions (using self-referential sentences, for example) that are of incomparable consistency strengths. By "natural" we mean considered by set theorists, because they had some set-theoretic idea behind them. Here the standards are very liberal, as the many thousands of pages published by set theorists will testify. There is some vagueness in "large cardinal hypothesis" too, but it is less important. Practically speaking, we may as well take "large cardinal hypothesis" to mean "informative marker of consistency strength, used to compare consistency strengths." Perhaps the main thrust of this vague conjecture at the moment is programmatic: understand better, and develop further, our methods for comparing consistency strengths. At present, this devolves at once into: understand better, and develop further, the theory of canonical inner models satisfying large cardinal hypotheses. One very ambitious conjecture here is that Con(PFA) $\Rightarrow$ Con(there is a supercompact cardinal). This has been a target of inner model theory for about 30 years.[9]

---

[8] The ones we know.

[9] There are various ways to attach ordinals to set theories that correspond to the consistency strength order in the case of natural theories. One can look at the provably recursive ordinals, or the minimal ordinal height of a transitive model, for example.

3   A theory of the concrete

A set theory T is consistent just in the case that all its Π_1^0 consequences are true. As emphasized by Kreisel (1983), this is the best way to understand Hilbert's program: a consistency proof for classical mathematics would justify the use of its ideal machinery in deriving "real," that is Π_1^0, statements. The consistency proof would yield a general elimination of classical mathematics from the proofs of Π_1^0 statements, in favor of whatever was used in the consistency proof.

Of course, Hilbert's dream that mathematicians of his time had found a final theory went up in smoke, and it is Gödel's program that rises from the ashes. But perhaps Hilbert's view that Π_1^0 sentences have a special status has something to it. They are, after all, the logically simplest, most concrete sentences that can be true, but not provable in ZFC. Remarkably, climbing the consistency strength hierarchy in any natural way seems to decide uniquely not just Π_1^0 sentences, but more complicated sentences about the concrete as well. *Concrete* refers here to natural numbers, real numbers, and certain sets of real numbers.

3.1   A theory of natural numbers

Definition 3.1 *Let* Γ *be a set of sentences in the syntax of* LST, *and* T *a set theory, then*

$$\Gamma_T = \{\varphi \mid \varphi \in \Gamma \wedge T \vdash \varphi\}.$$

We shall use this notation with $\Gamma = \Gamma = \Pi_1^0$, the set of Π_1^0 sentences, as well as $\Gamma = \Pi_\omega^0$ and $\Gamma = \Pi_\omega^1$, the sentences in the languages of first-order and second-order arithmetic, respectively.

It is not quite true that if $T \leq_{\mathrm{Con}} U$, then $\left(\Pi_1^0\right)_T \subseteq \left(\Pi_1^0\right)_U$. (We get a counterexample by letting T be U together with a Rosser sentence for U.) But it is true for natural theories,[10] and in fact we observe the following.

Phenomenon If T and U are natural extensions of ZFC, then

---

[10] Let $T \leq_{\mathrm{Con}}{}^* U$ iff for every finite $F \subset T$, U proves $\mathrm{Con}(F)$. By the reflection schema, for any S extending ZFC, $\left(\Pi_1^0\right)_S$ is axiomatized by $\{\mathrm{Con}(G) \mid G \subseteq S \wedge G \text{ is finite}\}$. Using this, it is easy to see that if T and U extend ZFC, then $T \leq_{\mathrm{Con}}{}^* U$ iff $\left(\Pi_1^0\right)_S \subseteq \left(\Pi_1^0\right)_U$. Finally, for natural theories extending ZFC, $T \leq_{\mathrm{Con}} U$ iff $T \leq_{\mathrm{Con}}{}^* U$.

$$T \leq_{\text{Con}} U \Leftrightarrow \left(\Pi^0_1\right)_T \subseteq \left(\Pi^0_1\right)_U$$
$$\Leftrightarrow \left(\Pi^0_\omega\right)_T \subseteq \left(\Pi^0_\omega\right)_U.$$

Thus the well ordering of natural consistency strengths corresponds to a well ordering by inclusion of theories of the natural numbers. There is no divergence at the arithmetic level, if one climbs the consistency strength hierarchy in any natural way we know of. Even more, the theory of the natural numbers generated by T is a monotonically increasing function of the consistency strength of T.[11]

3.2 A theory of the reals

Natural ways of climbing the consistency strength hierarchy do not diverge in their consequences for the reals.

Phenomenon Let T, U be natural theories of consistency strength at least that of "there are infinitely many Woodin cardinals"; then either

$$\left(\Pi^1_\omega\right)_T \subseteq \left(\Pi^1_\omega\right)_U \text{ or } \left(\Pi^1_\omega\right)_U \subseteq \left(\Pi^1_\omega\right)_T.$$

In other words, the second-order arithmetic generated by natural theories is an eventually monotonically increasing function of their consistency strengths.

This phenomenon extends to statements about sets of reals generated by reasonably simple means. As an example of the latter, we consider sets of reals in $L(\mathbb{R})$, the minimal model of ZF containing all reals and ordinals. Let us write $(\text{Th}^{L(\mathbb{R})})_T$ for the set of consequences of T of the form "$L(\mathbb{R}) \models \varphi$."

Phenomenon Let T, U be natural theories of consistency strength at least that of ZFC + "there are infinitely many Woodin cardinals with a measurable cardinal above them all"; then either

$$\left(\text{Th}^{L(\mathbb{R})}\right)_T \subseteq \left(\text{Th}^{L(\mathbb{R})}\right)_U \text{ or } \left(\text{Th}^{L(\mathbb{R})}\right)_U \subseteq \left(\text{Th}^{L(\mathbb{R})}\right)_T.$$

So there is no divergence at the level of statements about $L(\mathbb{R})$, if one climbs sufficiently high in the consistency strength hierarchy in any

---

[11] There are counterexamples to eventual monotonicity on the margins. For example, let T be ZFC + "there is an inaccessible cardinal," and U be ZFC + Con(T); then $T \leq_{\text{Con}} U$, but $\left(\Pi^0_\omega\right)_T \nsubseteq \left(\Pi^0_\omega\right)_U$. One could argue about whether U is natural. (Its extra axiom is about theories, not sets. It is an instrumentalist's theory.) In any case, $\left(\Pi^0_\omega\right)_T \cup \left(\Pi^0_\omega\right)_U \subseteq \left(\Pi^0_\omega\right)_S$, where S asserts that there are two inaccessibles. We do not know of an example of divergence, or even nondirectedness, with any claim to involve natural theories, in the realm of arithmetic.

natural way we know of. Even more, the theory of $L(\mathbb{R})$ generated by T is an eventually monotonically increasing function of the consistency strength of T.

The eventual monotonicity phenomenon extends somewhat beyond statements about $L(\mathbb{R})$. In general, we classify the complexity of statements by the type of objects they quantify over, and then by the number of alternations of such quantifiers. For example, statements of the form $V_\omega \models \varphi$ would be Σ_n^0, for some n. They quantify over type 0 objects (elements of V_ω, or up to simple coding, natural numbers). Statements of the form $V_{\omega+1} \models \varphi$ would be Σ_n^1, for some n. They quantify over type 1 objects (up to simple coding, real numbers). Statements of the form $V_{\omega+2} \models \varphi$ would be Σ_n^2, for some n. They quantify over type 2 objects (sets of reals). In between the Σ_n^1 and Σ_n^2 statements are statements about the reals with infinitely many alternations of quantifier. One can make precise sense of this by interpreting such statements using infinite games; see for example Moschovakis (2009). For example, if there is a measurable cardinal, statements of the form

$$L(\mathbb{R}) \models \varphi$$

can be expressed using the real game quantifier of length ω in the form

$$\exists x_0 \forall x_1 \exists x_2 \forall x_3 \ldots . (V_{\omega+1}, \in) \models \varphi[\vec{x}].$$

(The quantifier string is interpreted as saying that player I has a winning strategy in the game in which he plays the x_{2i} and his opponent plays the x_{2i+1}, and his goal is to insure that $(V_{\omega+1}, \in) \models \varphi[\vec{x}]|.$) One can obtain still larger classes of statements by considering games on the reals of length $> \omega$, and the eventual monotonicity phenomenon extends to some of these larger classes.

3.3   Completeness, correctness, and generic absoluteness

In addition to the question of divergence in the realm of the concrete, there is the question of completeness in this realm. Of course, no axiomatizable theory is literally complete, even in the realm of Π_1^0 statements, but we can ask for completeness with respect to mathematically natural statements. Here we find that, roughly speaking, if we climb to a consistency strength sufficient to insure monotonicity in the realm of Γ statements, then our theory decides the natural statements in the realm of Γ statements.

There is a partial explanation of the phenomena of nondivergence, eventual monotonicity, and practical completeness in the realm of the concrete, for theories of sufficiently high consistency strength. It lies in the way we obtain independence theorems, by interpreting one theory in another. Recall that there are two interpretation methods, one producing generic extensions, and one producing canonical inner models.

For natural T, U extending ZFC, if $T \leq_{Con} U$, then for every finite $F \subseteq T$, U proves "there is an ω-model M_F of F." This implies at once that $\left(\Pi_\omega^0\right)_T \subseteq \left(\Pi_\omega^0\right)_U$: if φ is an arithmetic consequence of T, then it is a consequence of some finite $F \subseteq T$, so it is true in M_F, so it is true. This is a partial explanation of arithmetic monotonicity, and it extends to Σ_2^1 monotonicity because, in practice, M_F will be a transitive model containing ω_1, and hence Σ_2^1-correct. If T and U are strictly stronger than "there are infintely many Woodin cardinals," then M_F will be $L(\mathbb{R})$-correct. In general, our model producing methods lead to eventual Γ-monotonicity because in order to produce a model for a theory T that is sufficiently strong with respect to Γ, we must produce a Γ-correct model.

Γ-nondivergence also provides a partial explanation of practical Γ-completeness, for natural T that are of sufficient consistency strength. For if φ is a Γ statement, and we obtained (by forcing or inner model theory) models M and N of T from models P and Q of the natural theories S and U, then for all we know, P and Q agreed on φ, and therefore M and N agreed on φ, because M and N were Γ-correct from the point of view of P and Q respectively.

Generic absoluteness theorems help explain why forcing yields Γ-correct models. Here are two such theorems at the level of statements about $L(\mathbb{R})$.

Theorem 3.2 (Woodin 1985) *Suppose there are arbitrarily large Woodin cardinals, and let M and N be set-generic extensions of V; then M and N satisfy the same statements about $L(\mathbb{R})$.*

Theorem 3.3 (Woodin 1988) *Suppose there is an iterable proper class model with infinitely many Woodin cardinals, and let M and N be set-generic extensions of V; then M and N satisfy the same statements about $L(\mathbb{R})$.*

The hypothesis of the second theorem follows from that of the first, so the second theorem properly contains the first. The second theorem also helps explain why canonical inner models of sufficiently strong large cardinal hypotheses are correct about the theory of $L(\mathbb{R})$.

The currrent large cardinal hypotheses are themselves generically absolute. If κ is measurable in V, then it is measurable in $V[G]$, whenever G is generic over V for a poset of size $<\kappa$. The same goes for Woodinness, superstrongness, supercompactness, etc. Large cardinal hypotheses resembling the current ones in very basic ways will be *preserved by small forcing*. If our hypothesis says there are arbitrarily large cardinals of some type, it will be preserved by set forcing. These are results of Lévy and Solovay from the 1960s.

Thus all the consequences of "there are arbitrarily large Φ-cardinals" are preserved by set forcing, for any of the current large cardinal properties Φ. Theorem 3.2 gives an indication of how extensive those consequences are, in one case.

Finally, we should point out one very important reason natural theories of sufficient consistency strength are complete in the realm of Γ statements: they imply Γ determinacy. Γ determinacy in turn yields a practically complete theory in the Γ realm. This theory coheres well with the theory of Borel and analytic sets developed by classical descriptive set theorists under ZFC, and in particular it bears out their intuition that definable sets of reals are free of the pathologies generated by a well order of the reals. Decades of work by modern descriptive set theorists have shown this. Since this key point has been made so often before, we shall not dwell on it here (see Martin 1998; Steel 2000; Moschovakis 2009). To give a concrete example: if there are arbitrarily large Woodin cardinals, then AD holds in $L(\mathbb{R})$, and this in turn is the basis for a thorough, detailed theory of the internal structure of $L(\mathbb{R})$.

4 A boundary

If we regard Theorem 3.2 as a key indicator of the completeness of large cardinal axioms for sentences about $L(\mathbb{R})$, then it is natural to ask whether this kind of completeness extends further. We can get near the boundary of what is known using the real game quantifier corresponding to games of length ω_1 which are *closed-Π_1^1*, in the sense that there is a Π_1^1 set A of (codes for) countable sequences of reals such that player I wins a run $\vec{x}$ just in the case that for all $\alpha < \omega_1$, $\vec{x} \restriction \alpha \in A$. Let ∂^{ω_1} (closed-Π_1^1) be the class of all statements of the form: player I wins such a game.

Theorem 4.1 (Woodin 1985) *Suppose there are arbitrarily large measurable Woodin cardinals, and let M and N be set-generic extensions of V; then M and N satisfy the same ∂^{ω_1} (closed-Π_1^1) statements.*

Theorem 4.2 (Woodin, Steel) *Suppose there is an iterable inner model satisfying "there is a measurable Woodin cardinal," and let M and N be set-generic extensions of V; M and N satisfy the same ∂^{ω_1} (closed-Π^1_1) statements.*

Whilst proving Theorem 4.1, Woodin also showed that if there is a measurable Woodin cardinal, then all length-ω games on ω with ∂^{ω_1} (closed-Π^1_1) payoff are determined. Recently, Itay Neeman has shown that the determinacy of the length-ω_1 closed-Π^1_1 games themselves follows from the existence of an iterable inner model satisfying "there is a measurable Woodin cardinal" (see Neeman 2007). The determinacy of these long games implies at once all length-ω games on ω with ∂^{ω_1} (closed-Π^1_1) payoff are determined. Neeman's theorem also gives a new proof of Theorem 4.2. In contrast to the situation lower down, we do not know yet that any large cardinal hypothesis implies that there is an iterable inner model satisfying "there is a measurable Woodin cardinal." Presumably, the existence of arbitrarily large measurable Woodins does so, and thus Neeman's determinacy theorem extends Woodin's, and yields proofs of both Theorem 4.2 and Theorem 4.1. But we have as yet no proof, and finding such a proof is a manifestation of the fundamental iterability problem in inner model theory.

What happens if we move from long strings of quantifiers over reals to quantifiers over sets of reals? Then we immediately encounter CH, which is a Σ^2_1 statement. None of our current large cardinal axioms decide CH, because they are preserved by small forcing, whilst CH can both be made true and be made false by small forcing.[12] Because CH is provably not generically absolute, it cannot be decided by large cardinal hypotheses that are themselves generically absolute.

Theorem 4.3 (Levy, Solovay) *Let A be one of the current large cardinal axioms, and suppose $V \models A$; then there are set-generic extensions M and N of V which satisfy $A + CH$ and $A + \neg CH$ repectively.*

There may be consistency strengths beyond those we have reached to date that cannot be reached in any natural way without deciding CH. But the consistency strengths of which we currently have some understanding can be reached by the current large cardinal axioms, and since these are generically absolute, they do not decide CH. On the other hand, the statement of CH involves just one quantifier over arbitrary sets of reals.

It is worth remarking that CH plays a fundamental role among Σ^2_1 sentences, in that it implies that every Σ^2_1 sentence can be translated into

---

[12] To vary CH, we need a poset of size at most $(2^{\aleph_0})^+$, so small with respect to any large cardinal.

a ∂^{ω_1} (closed-Π_1^1) sentence. Thus Theorem 3.3 gives us at once the following *conditional* generic absoluteness theorem for Σ_1^2 sentences.

Theorem 4.4 (Woodin) *Suppose $V \models$ "There are arbitrarily large measurable Woodin cardinals." Let M and N be set-generic extensions of V satisfying CH; then M and N are Σ_1^2-equivalent.*

It is open whether if $V \models$ "There are arbitrarily large supercompact cardinals," and M and N are set-generic extensions of V satisfying $\lozenge$, then M and N are Σ_2^2-equivalent. If so, this would indicate that in the presence of large cardinal hypotheses, $\lozenge$ yields a complete theory at the Σ_2^2 level, just as CH does at the Σ_1^2 level. Such *conditional generic absoluteness* theorems may go much further.

5 The multiverse language

What does this picture of what is possible suggest about what we should believe, or give preferred development, as a framework theory?

We have good evidence that the consistency hierarchy is not a mirage, that the theories in it we have identified are indeed consistent. This is especially true at the lower levels, where we already have canonical inner models, and equiconsistencies with fragments of definable determinacy. This argues for developing the theories in this hierarchy. All their Π_1^0 consequences are true, and we know of no other way to produce new Π_1^0 truths.

Developing one natural theory develops them all, via the Boolean-valued interpretations. At the level of statements about the concrete (including most of what non-set-theorists say), all the natural theories agree. This might suggest that we need no further framework: why not simply develop all the natural theories in our hierarchy as tools for generating true statements about the concrete? Let 1000 flowers bloom! This is Hilbertism without the consistency proof, and with perhaps an enlarged class of "real" statements.

The problem with this watered-down Hilbertism is that we do not want everyone to have his own private mathematics. We want one framework theory, to be used by all, so that we can use each other's work. It is better for all our flowers to bloom in the same garden. If truly distinct frameworks emerged, the first order of business would be to unify them.

In fact, the different natural theories we have found in our hierarchy are not independent of one another. Their common theory of the concrete stems from logical relationships that go deeper, and are brought out in our relative consistency proofs. These logical relationships may suggest a unifying framework.

The central role of the theories axiomatized by large cardinal hypotheses argues for adding such hypotheses to our framework. The goal of our framework theory is to *maximize interpretative power*, to provide a language and theory in which all mathematics, of today and of the future so far as we can anticipate it today, can be developed. Maximizing interpretative power entails maximizing consistency strength, but it requires more, in that we want to be able to translate other theories/languages into our framework theory/language in such a way that we preserve their meaning. The way we interpret set theories today is to think of them as theories of inner models of generic extensions of models satisfying some large cardinal hypothesis, and this method has had amazing success. We do not seem to lose any meaning this way. It is natural then to build on this approach.

Nevertheless, large cardinal hypotheses like our current ones cannot decide CH, and so our theory of the concrete still has many different possible theoretical superstructures, some with CH, some with $\Diamond$, some with MM, some with $2^{\aleph_0}$ being real-valued measurable, and so on: all the behaviors that can hold in set-generic extensions of V, no matter what large cardinals exist. Before we try to decide whether some such theory is preferable to the others, let us try to find a neutral common ground on which to compare them. We seek a language in which all these theories can be unified, without bias toward any, in a way that exhibits their logical relationships, and shows clearly how they can be used together. That is, we want one neat package they all fit into.

To this end, we describe a *multiverse language*, and an open-ended *multiverse theory*, in an informal way. It is routine to formalize completely.

Multiverse language: usual syntax of set theory, with two sorts, for the *worlds* and for the *sets*.

Axioms of MV:

(1)$_\varphi$ φ^W, for every world W (for each axiom φ of ZFC).

(2)   (a)   Every world is a transitive proper class. An object is a set only in the case that it belongs to some world.

      (b)   If W is a world and $\mathbb{P} \in W$ is a poset, then there is a world of the form $W[G]$, where G is $\mathbb{P}$-generic over W.

      (c)   If U is a world, and $U = W[G]$, where G is $\mathbb{P}$-generic over W, then W is a world.

      (d)   (Amalgamation) If U and W are worlds, then there are G, H sets generic over them such that $W[G] = U[H]$.

It is a theorem of Laver and Woodin (cf. Woodin 2011) that there is a formula ψ of LST such that if $N = W[H]$ where H is N-generic for a set forcing P, then ψ defines W over N from P and H. This result must be used to formalize (2)(c) precisely; W must be defined via ψ inside U in the statement of (2)(c).

The natural way to get a model of MV is as follows. Let M be a transitive model of ZFC, and let G be M-generic for $Col(\omega, < OR^M)$. The worlds of the multiverse M^G are all those W such that

$$W[H] = M[G \restriction \alpha],$$

for some H set generic over W, and some $\alpha \in OR^M$. It follows from the result of Laver and Woodin that the full first-order theory of M^G is independent of G, and present in M, uniformly over all M. That is, there is a recursive translation function t such that for any sentence φ of the multiverse language such that whenever M is a model of ZFC and G is Col $(\omega, < OR^M)$-generic over M, then

$$M^G \models \varphi \;\Leftrightarrow\; M \models t(\varphi),$$

for all sentences φ of the multiverse language. $t(\varphi)$ just says "φ is true in some (equvalently all) multiverse(s) obtained from me."

Notice also that if W is a countable model of MV, then for any world $M \in W$, there is a G such that $W = M^G$. Thus assuming MV indicates then that we are using the multiverse language as a sublanguage of the standard one, in the way described above. Also, it is clear that if φ is any sentence in the multiverse language, then MV proves

$$\varphi \Leftrightarrow \text{for all worlds } M, t(\varphi)^M \Leftrightarrow \text{for some world } M, t(\varphi)^M.$$

Thus everything that can be said in the multiverse language can be said using just one world-quantifier.

One can add large cardinal hypotheses that are preserved by small forcing to MV as follows: given such a large cardinal hypothesis φ, we add "φ^W, for all worlds W" to MV. For example, we might add "(there is a Woodin cardinal)^W, for all worlds W." By standard preservation theorems for large cardinals, this implies that every world has arbitrarily large Woodin cardinals. The same goes for superstrongs, supercompacts, etc. By adding large cardinal hypotheses to MV this way, we get as theorems "for all worlds W, φ^W," for any φ in the theory of the concrete they generate.

Why have we included the worlds in M^G; why not define it so as to have only the sets? Then we have only a model of "every set is countable"! We have lost important information, the information-level of the worlds. Set theory cannot be formalized in the language of this structure.

Why have we not put more into the multiverse? Why not declare definable inner models of worlds to be worlds, or even declare the sets to be worlds in their own right? The answer is that they are already there, we can talk about them in the multiverse language already, and if we go too far in the direction of obscuring what we are now calling the worlds with other objects, we may lose information.[13] Our multiverse is an equivalence class of worlds under "has the same information." Definable inner models and sets may lose information, and we do not wish to obscure the original information level.

For the same reason, our multiverse does not include class-generic extensions of the worlds. There seems to be no way to do this without losing track of the information in what we are now regarding as the multiverse, no expanded multiverse whose theory might serve as a foundation.[14] We seem to lose interpretative power.

The multiverse language is a sublanguage of the standard one, under the translation indicated above. It is sufficiently expressive to state versions of the axioms of ZFC, and of the large cardinal hypotheses preserved by set forcing: we replace φ by "for all worlds W, φ^W". Clearly we cannot state the CH this way. The same goes for the many other statements about the uncountable which are sensitive to set forcing, no matter what large cardinals there may be. Whether there are traces of CH and these other sentences in the multiverse language is the issue we consider next.

One can think of the standard language as the multiverse language, together with a constant symbol $\dot{V}$ for a reference universe. Statements like CH are intended as statements about the reference universe. To what extent is this constant symbol meaningful? Does one lose anything by retreating to the superficially less expressive multiverse language? We distinguish three answers to this question.

Weak relativist thesis Every proposition that can be expressed in the standard LST can be expressed in the multiverse language.

It follows from the weak relativist thesis that the symbol $\dot{V}$, which produces the standard language when we add it to the multiverse language, only makes sense if we can define it in the multiverse language. The weak relativist thesis is silent on whether one can do that.

---

[13] An extreme here is Hamkins' multiverse in Gitman and Hamkins (2010), whose first-order theory, if it were formalized so as to have one, would probably be an arithmetic set.

[14] Amalgamation will fail if we start counting sets, definable inner models, or class-generic extensions as worlds.

At the other pole, we have the following.

Strong absolutist thesis " $\dot{V}$ " makes sense, and that sense is not expressible in the multiverse language.[15]

Finally, perhaps weak relativism and the absolutist's idea of a distinguished reference world can be combined, in that that there is an individual world that is definable in the multiverse language. An elementary forcing argument shows that if the multiverse has a definable world, then it has a unique definable world, and this world is included in all the others.[16] In this case, we call this unique world included in all others the *core* of the multiverse.

Weak absolutist thesis There are individual worlds that are definable in the multiverse language; that is, the multiverse has a core.

The strongest evidence for the weak relativist thesis is that the mathematical theory based on large cardinal hypotheses that we have produced to date can be naturally expressed in the multiverse language. Perhaps we lose something when we do that, some future mathematics built around an understanding of the symbol $\dot{V}$ that does not involve defining $\dot{V}$ in the multiverse language. But at the moment, it is hard to see what that is.[17]

One argument for strong absolutism is that the multiverse language is parasitic on the standard one. We have only been able to understand it by giving the translation function t.[18] By itself, this is not a very strong objection, as one could think of what we are doing as isolating the meaningful part of the standard language, the range of t, while trimming away the meaningless, in order to avoid pseudo-questions. After climbing our ladder, we throw it away, and from now on, MV can serve as our foundation. After all, people learning how to use LST are shown diagrams in the shape of the

---

[15] The strong absolutist thesis is not, strictly speaking, the negation of the weak relativist one.

[16] The author is grateful to Woodin for this observation.

[17] There is a passage in Woodin (2011) in which, making some translations from Woodin's language to ours that may not be valid, the weak relativist is challenged to come up with some mathematical truth that cannot be formalized in the multiverse language (see Woodin 2011, p. 26). This is odd indeed, since the weak relativist thesis is just that, if we stay within the language of set theory, this cannot be done.

[18] Set theorists sometimes overestimate the value of models of set theory as a tool for fixing meaning. What we do in the formal semantics (model theory) of set theory is make translations. You do not get a model of ZFC unless you start with one. Our description of M^G showed how to translate the multiverse language into the standard one, but it did not explain the meaning of the multiverse language in any other way. The meaning of a foundational language is determined by its use, and in particular the theory we develop within it. The multiverse sublanguage has been in use a long time, and all of set theory has been developed within it.

letter "V," and told stories about random collection being iterated over transfinite time, and supplied with other metaphors with features that will produce pseudo-questions if taken too seriously. Perhaps one ought to discard more of this than is at first apparent. The true test is whether we thereby lose the ability to formalize any mathematics.

Whatever one thinks of the semantic completeness of the multiverse language, it does bring the weak absolutist thesis to the fore, as a fundamental question. Because the multiverse language is a sublanguage of the standard one, this is a question for everyone. If the multiverse has a core, then surely it is important, whether it is the denotation of the absolutist's $\dot{V}$ or not! Indeed, if there is an inclusion-least world in the multiverse, why not use $\dot{H}$ to denote it,[19] and agree to retire $\dot{V}$ until we need it? The question as to whether the multiverse has a core is an important question for everyone, relativist or absolutist.

Neither MV nor its extensions by large cardinal hypotheses of the sort we currently understand decides whether there is a core to the multiverse, or the basic theory of this core if it exists (see Reitz 2007; Hamkins *et al.* 2007; Hamkins 2011). So what we have here is another basic question, like the CH, that large cardinals do not decide. But it is a different question, and its role in our search for a universal framework theory seems crucial.

There is some reason to hope for a positive answer. Hugh Woodin has recently proposed an axiom which

(a) implies the multiverse has a core,
(b) suggests an approach toward developing a detailed, systematic "fine-structure theory" for this core, and
(c) may be consistent with all our large cardinal hypotheses.

The new mathematics needed in order to turn (b) and (c) from promise into reality is formidable, but there is some reason for optimism. In the next two sections I shall describe this axiom, where it comes from, and why one might hope that it pins down our multiverse, without restricting it.[20]

There are other ways the multiverse may have important global structure, as indicated by conditional generic absoluteness theorems. Let us say that an axiomatizable theory T in LST is Γ-*complete* if and only if T is true in some world of the multiverse, and whenever M and N are worlds

---

[19] As we shall see below, there is a reason for the letter H.

[20] The author wishes to thank Hugh Woodin for his permission to discuss his axiom, and his unpublished results regarding it.

satisfying T, then they satisfy the same Γ sentences.[21] If the weak absolutist thesis is true, then "I am the core of my multiverse" is Γ-complete, where Γ is the set of all sentences in LST whatsoever. So too is "I am a generic extension of the core of my multiverse via the poset adding ω_2 Cohen reals." But there are other interesting examples of Γ completeness. For example, CH is Σ_1^2-complete, by Theorem 4.1. There may be axiomatizable T that are Σ_n^2-complete for all n, and yet do not imply the weak absolutist thesis.

Historical note. The author has thought off-and-on about the various theses above for a long time, and published some remarks on the role of the generic multiverse at the end of Steel (2004). Recently, Hamkins (2011) and Woodin (2011) have published papers on the multiverse idea. Hamkins's realization of the idea is quite far from ours, and as far as I can tell, does not lead to any candidate for a framework language and theory. Woodin's generic multiverse V_M is much closer, but it is not actually a model of MV, and it is not at all clear what its theory would be.[22] Neither Hamkins nor Woodin presented a language and a first-order theory in that language, both of which seem necessary for a true foundation. If one takes Woodin's statement of what he calls the *generic multiverse position*, namely that "truth is equal to truth in all universes in V_M,"[23] and then tries to find a language that expresses all truths and all falsehoods and nothing else, one will probably arrive at the multiverse language and MV. The generic multiverse position seems to correspond to what I am calling the weak relativist thesis. Woodin's paper makes some arguments against the generic multiverse position, based on the logical complexity of certain truth predicates, but those arguments do not seem valid to me.[24] In particular, the weak relativist thesis does not seem to have much in common with formalist doctrines.

---

[21] If we require this also for M and N being rank initial segments of worlds, we get completeness in the sense of Woodin's $\models \Omega$ relation. Woodin has conjectured that no T is Σ_3^2-complete in this stronger sense, see Woodin (2011).

[22] V_M does not satisfy amalgamation.   [23] Where $M = V$?

[24] For those familiar with the arguments: the problem is that the decision to stay within the multiverse language does not commit one to a view as to what the multiverse looks like. The "multiverse laws" do not follow from the weak relativist thesis. The argument that they do is based on truncating the worlds at their least Woodin cardinals. However, this leaves one with nothing, an unstructured collection of sets with no theory. The Ω-conjecture does not imply a paradoxical reduction of MV or its language to something simpler, because there is no simpler language or theory describing a "reduced multiverse" consisting of truncated worlds. There is no "rejection of the transfinite beyond $H(\delta_0^+)$" implicit in staying within the multiverse sublanguage; indeed, saying that every world has a Woodin cardinal automatically entails that every world has arbitrarily large Woodin cardinals.

6 *V* looks like the HOD of a model of AD

Recall that a set is *ordinal definable* (OD) if and only if it is definable over the universe of sets from ordinal parameters, and is *hereditarily ordinal definable* (HOD) only in the case that it and all members of its transitive closure are OD. Gödel first isolated HOD in the 1940s; see Gödel (1946). Myhill and Scott (1971) showed that if $M \models$ ZF, then $\mathrm{HOD}^M \models$ ZFC. Woodin's axiom says that *V* looks like HOD^M, for models *M* of the axiom of determinacy.

Here is a precise statement. It is not the strongest form of the axiom one could put down, but is simple, and gives the idea.

Axiom H For any sentence φ of LST: if φ is true, then for some $M = \mathrm{AD}^+ + V = L(P(\mathbb{R}))$ such that $\mathbb{R} \cup \mathrm{OR} \subseteq M$, $(HOD \cap V_\Theta)^M \models \varphi$.

Here AD^+ is a slight strengthening of AD, introduced by Woodin for technical reasons that need not concern us here. Axiom H is a schema, but one could restrict it to Σ_2 sentences without much loss. The schema is stated above in the standard language, but Woodin has shown that it implies that *V* is the core of its own multiverse. So one could state Axiom H in the multiverse language: the multiverse has a core, and it satisfies Axiom H.

Axiom H holds in $(\mathrm{HOD} \cap V_\Theta)^M$,[25] if *M* is a model of $\mathrm{AD}_\mathbb{R}$ plus "Θ is regular." It is thus consistent relative to fairly weak large cardinal hypotheses (See Sargsyan 2009). The hope is that Axiom H is consistent with all the large cardinal hypotheses, so that adopting it does not restrict interpretative power, whilst at the same time it yields a detailed fine-structure theory for *V*, removing the incompleteness that large cardinal hypotheses by themselves can never remove.[26] It is known that Axiom H implies the CH, and many instances of the GCH. Whether it implies the full GCH is a crucial open problem.

Once again, Axiom H can be stated in the multiverse language. The strong absolutist who believes that *V* does not satisfy CH must still face the question whether the multiverse has a core satisfying Axiom H. If he agrees that it does, then the argument between him and someone who accepts Axiom H as a strong absolutist seems to have little practical importance.

---

[25] Θ is the least ordinal not the surjective image of the reals.
[26] If they are preserved by small forcing.

7 Homogeneously Suslin sets of reals

In this section, I shall say something about where Axiom H came from, and why one might entertain such hopes for it. The technical prerequisites for following this material are significantly greater than what was needed to follow the chapter up to this point. For the reader who is not a set theorist, it would make sense to skip this section.

What makes a set of reals amenable to the techniques of descriptive set theory?[27] The key property is that of being ∞-*homogeneously Suslin, or* Hom$_\infty$ for short. We shall give the technical definition now, but the reader will not lose much by reading Hom$_\infty$ as "well-behaved" in the sequel.

Definition 7.1 *A set $A \subseteq \omega^\omega$ is* Hom$_\infty$ *iff for any κ, there is a continuous function associating to reals x direct limit systems $\langle \left(M_n^x, i_{n,m}^x \right) \mid n, m < \omega \rangle$ on ω^ω such that for all x,*

(a) $M_0^x = V$, *each M_n^x is a transitive model of* ZFC *closed under κ-sequences, and*

(b) $x \in A \Leftrightarrow \lim_n M_n^x$ *is well founded.*

The concept was abstracted in the early 1970s by Kechris and Martin, from Martin's 1968 proof of Π_1^1-determinacy. Hom$_\infty$ sets are determined, Lebesgue measurable, have the Baire property, and so on. The definition seems to capture what it is about sets of reals that makes them "well behaved."[28] In general, a given large cardinal hypothesis will show that sets of reals in a certain logical complexity class are well behaved by showing that the sets in question are Hom$_\infty$.

A *pointclass* is a collection of sets of reals closed under complements and continuous pre-images. One can think of a pointclass as a logical complexity class for sets of reals. If there are arbitrarily large Woodin cardinals, then Hom$_\infty$ is a pointclass, and closed under fairly complicated types of definability. In fact, we have the following.

Theorem 7.2 (S. Martin, Woodin 1985) *If there are arbitrarily large Woodin cardinals, then for any pointclass Γ properly contained in* Hom$_\infty$, *every set of reals in $L(\Gamma, \mathbb{R})$ is in* Hom$_\infty$, *and thus $L(\Gamma, \mathbb{R}) \models AD^+$.*

---

[27] For us, a real is usually an infinite sequence of natural numbers, considered as an element of the Baire space ω^ω.

[28] *Universally Baire* is an intensionally weaker property, discovered in the late 1980s by Feng *et al.* (1992). All Hom$_\infty$ sets are universally Baire, and if there are arbitrarily large Woodin cardinals, then all universally Baire sets are Hom$_\infty$. See Steel (2009).

Here AD^+ is a slight strengthening of the axiom of determinacy AD. The fact that $L(\Gamma, \mathbb{R}) \models AD^+$ leads to a thorough understanding of this model. One of the most basic features of such a model is given by the following theorem.

Theorem 7.3 (Wadge, Martin 1970) *Assume* AD; *then the pointclasses are well ordered by inclusion.*

By Theorem 7.3, the sets of reals in Hom_∞ fall into a well-defined, well-ordered hierarchy based on their logical complexity. As we climb this hierarchy, new facts as to the existence of well-behaved sets of reals with various properties may be verified. More precisely, new $(\Sigma_1^2)^{Hom_\infty}$ statements may be verified, where a $(\Sigma_1^2)^{Hom_\infty}$ statement is one of the form: $\exists A \in Hom_\infty(V_{\omega+1}, \in, A) \models \varphi$.

Definition 7.4 0^Ω *is the set of all true* $(\Sigma_1^2)^{Hom_\infty}$ *statements. For any theory* T, *put* $0_T^\Omega = \{\varphi \mid \varphi \text{ is } (\Sigma_1^2)^{Hom_\infty} \text{ and } T \vdash \varphi\}$.

Phenomenon Let T, U be natural theories extending ZFC + "there are arbitrarily large Woodin cardinals"; then either $0_T^\Omega \subseteq 0_U^\Omega$ or $0_U^\Omega \subseteq 0_T^\Omega$.

There is a version of this phenomenon that applies to theories that are consistencywise as strong as "there are arbitrarily large Woodin cardinals," rather than outright extend that theory, but it would take us too far afield to try to explain it.

7.1 Generic absoluteness revisited

The central idea of descriptive set theory is that the *definable* sets of reals are well behaved (Lebesgue measurable, determined, etc.) in ways that arbitrary sets of reals are not. How is definability related to homogeneity? In brief, a large cardinal hypothesis H will prove that the pointclass Γ of sets of reals definable in a certain way consists of well-behaved sets by showing $\Gamma \subseteq Hom_\infty$.

In a similar vein, the generic absoluteness of a class of statements can be proved by reducing them to $(\Sigma_1^2)^{Hom_\infty}$ statements. The following theorem is an abstract statement of part of the method (see Steel 2009).

Theorem 7.5 (Woodin) *If there are arbitrarily large Woodin cardinals, then* $(\Sigma_1^2)^{Hom_\infty}$ *statements are absolute for set forcing.*

In order to prove a class Γ of statements is generically absolute under the large cardinal hypothesis H this way, one produces a recursive translation function t, and uses H to show that for all φ,

$$\varphi \text{ is true} \Leftrightarrow t(\varphi) \in 0^{\Omega}.$$

One may need an H stronger than "there are arbitrarily large Woodin cardinals" to do this. For example, if Γ is the class of ∂^{ω_1} (closed-Π_1^1) statements, then H must be "there is an iterable model with a measurable Woodin cardinal." It is natural to view the proof of Theorem 4.2 this way.

The proofs of generic absoluteness that proceed directly from large cardinals, such as that of Theorem 4.1, need not produce reductions of Γ-truth to $(\Sigma_1^2)^{\text{Hom}_\infty}$-truth. However, it is reasonable to believe that one day the iterability problem will be solved, and that Γ-generic absoluteness will be shown to follow from reductions of Γ-truth to $(\Sigma_1^2)^{\text{Hom}_\infty}$-truth, for very complicated Γ.

How complicated? It is easy to see that every $(\Sigma_1^2)^{\text{Hom}_\infty}$ sentence φ is (up to provable-in-ZFC equivalence) a Π_2 sentence of LST.[29] Assuming there are arbitraily large Woodin cardinals, φ is therefore equivalent to the sentence in the multiverse sublanguage: "for all worlds W, φ^{W}". This puts an upper bound on the complexity of sentences whose generic absoluteness can be guaranteed by a reduction to 0^{Ω}.

Woodin's Ω-conjecture (cf. Woodin 2011) implies that this upper bound is optimal. It states that any true sentence of the multiverse sublanguage of the form "for all worlds W, φ^{W}", where φ is Π_2, has a Hom$_\infty$ proof, in a certain precisely defined sense. If the Ω-conjecture is true, then generic absoluteness for Π_2 statements is always guaranteed by Hom$_\infty$ proofs, in a way that resembles abstractly the proofs of Theorems 3.3 and 4.2. So up to provable equivalence in ZFC + "there are arbitrarily large Woodin cardinals," the class of $(\Sigma_1^2)^{\text{Hom}_\infty}$ statements coincides with the class of statements of the form "for all worlds W, φ^{W}", where φ is Π_2.

It is worth pointing out that the large cardinal hypotheses are Σ_2, and sometimes Σ_3. (The "local" ones like the existence of measurables, Woodins, superstrongs, or huges are Σ_2. The existence of strong or supercompact cardinals is a Σ_3 sentence.) "For all worlds W (there is a measurable)^{W}" is a true statement of the multiverse language. This makes

---

[29] The best way to think of Π_2 sentences is to think of them as Π_1 relative to the power set function. In other words, they are just sentences of the form "for all α, $V_\alpha \models \psi$."

it hard to see how the weak relativist thesis implies an identification of truth with Ω-provability, as Woodin (2011) seems to argue.[30]

Here are two natural test questions regarding the extent of generic absoluteness.

Open questions Does any large cardinal hypothesis (e.g., the existence of arbitrarily large supercompact cardinals) imply

(a) that statements of the form $\forall x \in \mathbb{R} \exists A \in \text{Hom}_\infty \, (V_{\omega+1}, \in, A) \models \varphi[x]$ are absolute for set forcing?
(b) that $L(\text{Hom}_\infty, \mathbb{R}) \models AD$?

The canonical inner models for such a large cardinal hypothesis would have to be different in basic ways from those we know. It is unlikely that superstrong cardinals would suffice.

Axiom H plus "there are arbitrarily large Woodin cardinals" implies the negations of both (a) and (b).

7.2   Extender models and iteration strategies

The natural attempt to complete set theory in a way that is compatible with all the large cardinal hypotheses is to try to model those hypotheses in some minimal way. This leads us into the theory of canonical inner models for large cardinal hypotheses. Gödel's L is the prototypical such model, but it is too small to satisfy even moderately strong large cardinal hypotheses. In the period 1966–1990, set theorists generalized Gödel's construction in such a way that it could produce models with superstrong cardinals. Let us call the resulting models *short-extender models*.[31] It is not known how to generalize this framework so that it produces inner models with supercompact cardinals and beyond. The author believes this can be done. Let us call the as-yet-undiscovered models *long-extender models*.

What makes a countable short-extender model M canonical is a Hom_∞ set of reals called an *iteration strategy for M*. Let us call such M *iterable*. If M is iterable, then every real in M is ordinal definable, and in fact, $(\Sigma_1^2)^{\text{Hom}_\infty}$ -definable from a countable ordinal. Moreover, if φ is a

---

[30] Gödel sentences like "I have no Hom_∞ proof from T" and "there is no Hom_∞ proof of $0 = 1$ in T" also show that truth for sentences of the form "for all W, φ^W," with φ being Σ_2, exceeds Hom_∞-provability from T. Here we assume T is an axiomatizable theory extending ZFC+ "there are arbitrarily large Woodin cardinals," and T is true in V_α^W, for some world W and some ordinal α.
[31] The basics of this framework are due to Mitchell (1979).

$(\Sigma_1^2)^{\text{Hom}_\infty}$ statement, and there is an iterable short-extender model M with arbitrarily large Woodin cardinals such that $M \models \varphi$, then φ is true. The converse is true for φ that are not too complicated. For example, true $(\Sigma_1^2)^{\text{Hom}_\infty}$ statements of the form $(V_{\omega+1}, \in, \mathbb{R}^\#) \models \varphi$ are certified by iterable short-extender models this way, and this is the basis for the proof of Theorem 3.3. More generally, the Hom_∞ proofs of Ω-logic, in the region where we understand them very well, are Hom_∞ iteration strategies for short-extender models.

It is natural to guess that the as-yet-undiscovered long-extender models are also certified by Hom_∞ sets, and that a $(\Sigma_1^2)^{\text{Hom}_\infty}$ statement is true just in case it is true in some iterable extender model. That is, the Hom_∞ proofs of Ω-logic can always be taken to be an iteration strategy for a canonical extender model reaching some large cardinal hypothesis or other. If this is not the case, then Axiom H loses plausibility, for the determinacy models in which those HODs are formed may just not go far enough. We are still some considerable distance from proving this general *mouse set conjecture* (see Steel and Woodin 2012). We do not know that there are iteration strategies for short-extender models with superstrongs, and we do not know what the long-extender models look like (Woodin (2010b) has made progress on the latter problem).

7.3   HOD in models of determinacy

There is another family of canonical models of ZFC. Let $M = L(\Gamma, \mathbb{R})$, where Γ is a pointclass properly included in Hom_∞. Put

$$\mathcal{H}^M = \text{HOD}^M \cap V_\theta,$$

where $\theta = \Theta^M$. Note that $\mathcal{H}^M$ really only depends on the Wadge ordinal of Γ, and therefore it is contained in the HOD of V. Indeed, all its reals are $(\Sigma_1^2)^{\text{Hom}\infty}$ in a countable ordinal.

To simplify the discussion a bit, let us assume M is a model $\text{AD}_\mathbb{R} +$ "Θ is regular." By results of Sargsyan (2009, the existence of a Woodin limit of Woodin cardinals implies that there are such M. Hugh Woodin has shown that in this case

$$\mathcal{H}^M \models \text{ZFC} + \text{``there are arbitrarily large Woodin cardinals''}$$

and that every $(\Sigma_1^2)^{\text{Hom}_\infty}$ truth with witness in Γ is true in $\mathcal{H}^M$. (See Koellner and Woodin (2010) for the first statement, and Steel (2009) for a proof of the second.) So $\mathcal{H}^M$ keeps pace with M with respect to $(\Sigma_1^2)^{\text{Hom}_\infty}$ truth, and it has at least moderately large cardinals. Moreover, Woodin has shown that

$$\mathcal{H}^M \models \text{Axiom H.}$$

(For the experts: $\mathcal{H}^M$ can be elementarily embedded into the HOD of its own derived model. But this derived model is a proper Wadge initial segment of $j((\text{Hom}_\infty)^{\mathcal{H}^M})$, for some stationary tower embedding j. So $\mathcal{H}^M$ is elementarily equivalent to the HOD of some proper initial segment of $j((\text{Hom}_\infty)^{\mathcal{H}^M})$, and now we can pull this fact back using j.)

Can $\mathcal{H}^M$ be analyzed? It is known to satisfy CH, and many instances of GCH,[32] but in general we do not know. However, Sargsyan has shown that for M_0 the minimal Wadge initial segment of Hom_∞ satisfying $\text{AD}_{\mathbb{R}} +$"Θ is regular," a full analysis is possible. Sargsyan's analysis actually applies to the HOD of $L(\Gamma, \mathbb{R})$, for any Wadge initial segment of Hom_∞ up to, and somewhat past, M_0.[33] For the M to which Sargsyan's work applies, HOD^M is a short-extender model, expanded with information about what its own iteration strategy is.[34] It is natural to suppose that this analysis will go all the way, and that one day we will prove the following:

Conjecture If Γ is a pointclass properly contained in Hom_∞, then GCH holds in $\text{HOD}^{L(\Gamma, \mathbb{R})}$.

The conjecture is a $(\Pi_1^2)^{\text{Hom}_\infty}$ statement, so it is generically absolute. We would guess that it is provable in ZFC + "there are arbitrarily large Woodin cardinals."

The other important question here is whether $\mathcal{H}^M$ can satisfy very strong large cardinal hypotheses. We do not know yet whether even a Woodin limit of Woodins is possible.[35] If $M = \text{AD}_{\mathbb{R}}$, then $\mathcal{H}^M$ has no strong cardinals, by a theorem of Woodin. However, the local, Σ_2 large cardinal axioms may hold in $\mathcal{H}^M$ for sufficiently large $M \models \text{AD}_{\mathbb{R}}$, and the global ones, like the existence of supercompacts, may hold in $\mathcal{H}^M$ for M that do not satisfy $\text{AD}_{\mathbb{R}}$.[36]

---

[32] Woodin's proof that Σ_1^2 has the scale property (cf. Steel 2009) implies that for $\Gamma = \Sigma_1^2$, the results of Moschovakis (1981) apply.

[33] Sargsyan's work builds on work of the author and Woodin (see Steel and Woodin 2012).

[34] The idea of expanding extender models this way is due to Woodin.

[35] Sargsyan contributed to Woodin's discovery of Axiom H by persisting in the idea that this might be possible.

[36] Which of these two kinds of HOD is a model for the core of the multiverse seems to me to be a matter of convention, like the decision between Kelley–Morse and ZFC. Indeed, the universe above a strong cardinal is something like the level of proper classes in Kelley–Morse.

8 Gödel's program

What are our prospects today for reaching Gödel's original goal, deciding the CH?

The work of Gödel, Cohen, and their successors has shown us just how important the metamathematics of set theory is in this endeavor. An understanding of the possible set theories is useful in finding the true theory. Our metamathematical work has also shown us just how many other natural questions are in the same boat as the CH, and thereby broadened our focus. It seems highly unlikely that the solution to CH will be a "one-off."[37] One idea motivating Gödel's work on L, the first substantial work in the metamathematics of set theory, was that one cannot count the reals until one has specified in a more thorough way what it is one is counting. This seems more likely to carry the day. Most likely, the continuum problem will not be solved without a significant, far-reaching clarification of the notion of set. Our metamathematical work is a necessary prelude to that.

Our current understanding of the possibilities for maximizing interpretative power has led us to one theory of the concrete, and a family of theoretical superstructures for it, each containing all the large cardinal hypotheses. These different theories are logically related in a way that enables us to use them all together. Whatever the strong absolutist may believe about V, it is surely an important fact about the global structure of V that it has the generic extensions it does have.

The logical relationships between the different theories extending ZFC plus large cardinal hypotheses we have discovered are brought out clearly by formalizing them in the multiverse language. This language is a sublanguage of the standard one, and in it we can formalize naturally all the mathematics that set theorists have done. Remaining within this sublanguage has the additional virtue that our attention is directed away from CH, which has no obvious formalization within it, and toward the global question as to whether the multiverse has a core.

Remarkably, we can see now the outlines of a positive answer to this question, a way in which the multiverse may indeed have a core, and this core may admit a detailed fine-structural analysis that resembles that of Gödel's L. There are formidable technical mathematical problems that

---

[37] Freiling's "darts argument" and some passages in Gödel (1947) are examples of attempted "one-offs." The Banach–Tarski paradox ought to make clear how little weight can be attached to appeals to intuition in this context.

need to be answered in a certain way to realize this promise: we must show that there are models M of AD^+ such that HOD^M satisfies "there are supercompact cardinals," for example, and we must produce a fine-structure theory for HOD^M. Although these are difficult questions, large cardinal hypotheses should settle them.

Perhaps the mathematics will turn out some other way. Perhaps the multiverse has no core, but some other, more subtle structure. There are many basic open questions at the foundations of set theory: the extent of generic absoluteness, the existence of iterable structures, the Ω-conjecture, the form of canonical inner models with supercompacts, and the properties of HOD in models of determinacy, to give my own partial list. Our path toward a stronger foundation will be lit by the answers to such questions.

CHAPTER 9

Multiverse set theory and absolutely undecidable propositions

Jouko Väänänen

I Introduction

After the incompleteness theorems of Gödel, and especially after Cohen proved the independence of the continuum hypothesis from the ZFC axioms, the idea offered itself that there are *absolutely undecidable* propositions in mathematics, propositions that cannot be solved at all, by any means. If that were the case, one could throw doubt on the idea that mathematical propositions have a determined truth-value and that there is a unique well-determined reality of mathematical objects where such propositions are true or false. In this chapter we try to give this doubt rational content by formulating a position in the foundations of mathematics which allows for multiple realities, or "parallel universes."

The phrase "multiple realities," as well as "parallel universe," may sound immediately self-contradictory and ill defined. We try to make sense of it anyway. It helps perhaps to look forward: according to our concept of "multiple realities," a working set theorist will not be able to be sure whether there are multiple realities or just one,[1] and will certainly not be able to talk about individual realities.

The word "reality" is famously not unproblematic in the foundations of mathematics, but we are only concerned in this chapter with the question whether it makes sense to talk about multiple realities or not, assuming it makes sense to talk about reality at all. We are not concerned with the problem of what "reality" means, apart from the multiplicity question.

Let us perform a thought experiment[2] to the effect that there are two realities, or "parallel universes," in mathematics, V_1 and V_2. Suppose we

This research was partially supported by grant 40734 of the Academy of Finland and a University of Amsterdam–New York University exchange grant, which enabled the author to visit New York University Philosophy Department as a visiting scholar during the academic year 2011–2012.
[1] Unless he or she adopts the stronger language of Section 5.
[2] Familiar from the supervaluation theory of truth.

have a sentence φ, perhaps the continuum hypothesis itself, that is true in V_1 but false in V_2. Obviously we would not say that φ is *true*, because it is in fact false in V_2. Neither would we say that it is false, because it is true in V_1. So it is neither true nor false. What about its negation $\neg\varphi$? Since φ is not true, should we not declare $\neg\varphi$ true? But if $\neg\varphi$ is true, why is φ not declared false? If negation has lost its meaning, have we lost also faith in the Law of Excluded Middle $\varphi \vee \neg\varphi$? What has happened to the laws of logic in general?

The above thought experiment shows that allowing a divided reality may call for a re-evaluation of the basic logical operations and laws of logic. However, we can keep all the familiar laws of (classical) logic if we decide to call "true" those propositions that hold in *both* V_1 and V_2, and "false" those propositions that are false in *both* V_1 and V_2. A disjunction is called "true" if one disjunct is true in V_1 and the other is true in V_2. Thus $\varphi \vee \neg\varphi$ is still true, whatever φ, despite the fact that φ itself is neither true nor false. By developing this approach to the interpretation of logical constants, we can make sense of the situation that there are two realities. At the same time, we make sense of the situation that some propositions are absolutely undecidable: they are absolutely undecidable because they are true in one reality and false in another.

The reader will undoubtedly ask, is this not just what Gödel proved in his completeness theorem: undecidability of φ by given axioms ZFC means the existence of two models M_1 and M_2 of ZFC, one for φ and the other for $\neg\varphi$. This is indeed the "outside" view about a theory, such as ZFC, familiar already to Skolem (1923) and von Neumann (1925). But we are trying to make sense of this from the "inside" of ZFC. A theory like ZFC is a theory of all mathematics; everything is "inside" and we cannot make sense of the "outside" of the universe inside the theory ZFC itself, except in a metamathematical approach. If we formulate V_1 and V_2 inside ZFC in any reasonable way, modeling the fact that they are two "parallel" versions of V, it is hard to avoid the conclusion that $V_1 = V_2$, simply because V is "everything." This is why the working set theorist will not be able to recognize whether he or she has one or several universes.

Already von Neumann (1925) introduced the concept of an inner model and Gödel made this explicit in his universe L of constructible sets. If we assume the existence of a σ-complete total measure on the reals, we must conclude $V \neq L$ (Ulam). Do we not have in this case two universes, L and V? The difference with the above situation with V_1 and V_2 is that in the former case we *know* that L is not the entire universe, but in the latter case we consider both V_1 and V_2 as being the entire universe, whatever this means.

Our problem is now obvious: we want two universes in order to account for absolute undecidability and at the same time we want to say that both universes are "everything." We solve this problem by thinking of the domain of set theory as a *multiverse* of parallel universes, and letting variables of set theory range – intuitively – over each parallel universe simultaneously, *as if* the multiverse consisted of a Cartesian product of all of its parallel universes.[3] The axioms of the multiverse are just the usual ZFC axioms and everything that we can say about the multiverse is in harmony with the possibility that there is just one universe.[4] But at the same time the possibility of absolutely undecidable propositions keeps alive the possibility that, in fact, there are several universes. The intuition that this chapter is trying to follow is that the parallel universes are more or less close to each other and differ only "at the edges."

Our *multiverse* consists of a multitude of universes. Truth in the multiverse means truth in each universe separately, with the same meaning for falsity. Thus negation does not have the usual meaning of not-true. Still the Law of Excluded Middle, as well as other principles of classical logic, are valid. Absolutely undecidable propositions are true in some universes of the multiverse and false in some others. So an absolutely undecidable proposition is neither true nor false, i.e., it lacks a truth-value. The idea is *not* that every model that the axioms of set theory admit is a universe in the multiverse; that would mean that we could dispense with the multiverse entirely and talk only about the axioms.

We are not admitting[5] the possibility that mathematical propositions do have truth-values but for some of them mathematicians will never be able to figure out what the truth-value is. We are only concerned with the possibility that mathematicians are never able to find the truth-value of some proposition because such a truth-value does not exist.

It is the purpose of this chapter to present the multiverse approach to set theory in detail. In Section 2 we give some background and a review of views on absolute undecidability, of Gödel and others. In Section 3 we present the multiverse. In Section 4 we present elements of first-order logic in the multiverse setup. In Section 5 we introduce new methods, based on Väänänen (2007), to get a better understanding of the multiverse.

---

[3] But the Cartesian product is just a mental image. We cannot form the Cartesian product because we cannot even isolate the universes from each other.
[4] Until we start using the stronger methods of Section 5.
[5] Not only because the human race may by wiped out tomorrow.

2 Background

John von Neumann wrote in 1925:

> Das abzählbar Unendliche als solches ist unanfechtbar: es ist ja nichts weiter als der allgemeine Begriff der positive ganzen Zahl, auf dem die Mathematik beruht und von dem selbst Kronecker und Brouwer zugeben, daß er von "Gott geschaffen" sei. Aber seine Grenzen scheinen sehr verschwommen und ohne anschaulich-inhaltliche Bedeutung zu sein. (von Neumann 1925)[6]

An extreme form of the multiverse idea is the claim that there is no more truth in set theory than what the axioms give. Von Neumann writes:

> Unter "Menge" wird hier (im Sinne der axiomatischen Methode) nur ein Ding verstanden, von dem man nicht mehr weiß und nicht mehr wissen will, als aus den Postulaten über es folgt. (von Neumann 1925)[7]

Von Neumann refers to Skolem and to Löwenheim (1915) as sources of the noncategoricity of his, or any other, set theory. It is worth noting that von Neumann puts so much weight on categoricity. Indeed, if set theory had a categorical axiomatization, the categoricity proof itself, carried out in set theory, would be meaningful. But with noncategoricity everything is lost (see, however, Väänänen 2012).

For a time Gödel contemplated the idea that there could be *absolutely undecidable* propositions in mathematics.[8] He wrote:

> The consistency of the proposition A (that every set is constructible) is also of interest in its own right, especially because it is very plausible that with A one is dealing with an absolutely undecidable proposition, on which set theory bifurcates into two different systems, similar to Euclidean and non Euclidean geometry. (Gödel 1995, p. 155)

Later Gödel turned against this view:

> For if the meanings of the primitive terms of set theory as explained on page 262 and in footnote 14 are accepted as sound, it follows that the set-theoretical concepts and theorems describe some well-determined reality,

---

[6] "The denumerable infinite as such is beyond dispute; indeed, it is nothing more than the general notion of the positive integer, on which mathematics rests and of which even Kronecker and Brouwer admit that it was 'created by God'. But its boundaries seem to be quite blurred and to lack intuitive, substantive meaning." (English translation from van Heijenoort 1967a.)

[7] "Here (in the spirit of the axiomatic method) one understands by 'set' nothing but an object of which one knows no more and wants to know no more than what follows about it from the postulates."

[8] For more on Gödel's views on absolute undecidability see van Atten and Kennedy (2009).

in which Cantor's conjecture must be either true or false. (Gödel 1947, reprinted in Benacerraf and Putnam 1983, p. 260)

it has been suggested that, in case Cantor's continuum problem should turn out to be undecidable from the accepted axioms of set theory, the question of its truth would lose its meaning, exactly as the question of the truth of Euclid's fifth postulate by the proof of the consistency of non Euclidean geometry became meaningless for the mathematician. I therefore would like to point out that the situation in set theory is very different from that in geometry, both from the mathematical and from the epistemological point of view. (Benacerraf and Putnam 1983, p. 267)

We can study geometries in set theory, but not the other way around. More importantly, there is no stronger theory in which we would study set theory.[9] Set theory is meant to be the ultimate foundation for all mathematics. If we imagined a mathematical theory of models of set theory T, we would need a background theory in which this would be possible. If that background theory is a set theory T^*, we again must ask, is T^* talking about one universe or a multiverse? A lot of the investigation of set theory since Cohen's result on the continuum hypothesis (CH) can be seen as a study of models of finite parts of ZFC, but no stronger theory ZFC* is needed because ZFC can prove the existence of models for any of its finite parts. The goal, following Cohen's result, is not so much to show that reality has many facets but rather to show that the axioms leave many things undecided. Still the fact that so many things are left undecided lends credibility to the idea that this is not *only* because the axioms are too weak but also because they try to describe something which is not unique.

Another sense in which the independence of Euclid's Fifth Postulate is different from the independence of CH was pointed out by Kreisel (1967, 1(b)). The Fifth Postulate is undecided even from the second-order axioms of geometry, while second-order axioms in set theory fix the levels of the cumulative hierarchy (Zermelo 1930) and thereby fix CH. So the independence of CH is in this respect of a weaker kind than the independence of Euclid's Fifth Postulate.

Saharon Shelah has emphasized the interest in proving set-theoretical results in ZFC alone and has demontrated the possibilities with his pcf theory (Shelah 1994b). On the universe of set theory Shelah writes:

---

[9] Apart from class theories such as the Mostowski–Kelley–Morse impredicative class theory. But these do not change the basic questions.

I am in my heart a card-carrying Platonist seeing before my eyes the universe of sets, but I cannot discard the independence phenomena. (Shelah 1993b)

... I do not agree with the pure Platonic view that the interesting problems in set theory can be decided, we just have to discover the additional axiom. My mental picture is that we have many possible set theories, all conforming to ZFC. (Shelah 2003a)

3 The multiverse of sets

The informal description of the multiverse is very much like an informal description of the universe of sets. So we start with an overview of the one universe view.

3.1 The one universe case

The so-called *iterative* concept of set[10] soon became entrenched in set theory in the early twentieth century. Let us recall the basic idea. Roughly speaking, the universe of set theory is, according to the iterative set view, the closure of the urelements[11] (aka individuals) under iterations of the power-set operation and taking unions. The crucial factors are the power-set operation and the length of the iteration. It seems difficult to say what the power-set of an infinite set should be like, apart from being closed under rather obvious operations and containing subsets that are actually definable.

Satisfying the Axiom of Choice in the final universe requires us to add choice sets for sets of nonempty sets, and this is a potential source of variation. Different ways to choose the choice sets may lead to different universes. The one universe view holds that the choice-functions can be chosen in a canonical way, leading to a unique universe. Of course, no actual "selecting" takes place because the whole picture of iterative set is just a helpful image for understanding the axioms.

To make the iterative concept of set even more intuitive, the concept of a *stage* was introduced.[12] The concept of stage takes from the concept

---

[10] The iterative concept of set was first suggested by Mirimanoff (1917) and made explicit by von Neumann (1925). For a thorough discussion of this concept of set see Boolos (1971) and Parsons (1977).

[11] Quickly found unnecessary.

[12] Apparently this explanatory concept is folklore. It features in Shoenfield (1967) and again in Barwise (1977, pp. 321–344).

of iterative set the aspect of iteration: elements of a set are thought to have been formed at stages *prior to* the stage where the set itself is formed. As Shoenfield (1967) explains, "prior to" is not meant in a temporal sense but rather in a logical sense, as when we say that one theorem must be proved before another.

The idea of first focusing on the stages suggests itself naturally. If the stages are thought to be (intuitively) well ordered, one can rely on the strong rigidity of well-orders. When Gödel formed the inner model HOD of hereditarily ordinal definable sets he noted:

> ... in the ordinals there is certainly no element of randomness, and hence neither in sets defined in terms of them. This is particularly clear if you consider von Neumann's definition of ordinals, because it is not based on any well-ordering relations of sets, which may very well involve some random element. (Gödel 1946, reprinted in Gödel 1990)

Unlike ZFC set theory itself, the theory of well-order is decidable[13] and its complete extensions are well understood. This further emphasizes the advantage of taking the concept of a stage as a stepping stone in the understanding of the iterative concept of set. Indeed, Boolos (1971) formalizes the concept of *stage* as his *stage theory* and derives the ZFC axioms except the Axiom of Choice from that theory.

Although the universe is, according to the iterative set view, the *minimal* universe closed under the said operations and iterations, there is a commonly held view that the universe should be at the same time *maximal*, for otherwise we face immediately the question, what else is there, outside the universe so to speak.

There is no general agreement about what would be the right criterion of maximality. According to the independence results of Cohen, we can make CH true either by restricting the reals of the universe (to Gödel's L), or by adding reals to the universe (e.g., by starting with $2^{\aleph_0} = \aleph_2$ and collapsing $\aleph_1$ to $\aleph_0$). This demonstrates the basic problem of finding criteria for maximality.

One avenue to maximality, already emphasized by Gödel (1946)[14] is to maximize the length of the iteration by means of Axioms of Infinity, such as the assumption of inaccessible (and larger) so-called large cardinals.

Gödel seems to have strongly favored the iterative concept of set:

[13] Mostowski and Tarski (1949), proved in Doner (1978).
[14] "Remarks before the Princeton Bicentennial Conference on problems in mathematics," 1946, reprinted in Gödel (1990); see pp. 150–153.

As far as sets occur in mathematics (at least in the mathematics of today, including all of Cantor's set theory), they are sets of integers, or of rational numbers (i.e., of pairs of integers), or of real numbers (i.e., of sets of rational numbers), or of functions of real numbers (i.e., of sets of pairs of real numbers), etc. When theorems about all sets (or the existence of sets in general) are asserted, they can always be interpreted without any difficulty to mean that they hold for sets of integers as well as for sets of sets of integers, etc. (respectively, that there either exist sets of integers, or sets of sets of integers, or ... etc., which have the asserted property). (Gödel 1990, p. 258)

As Gödel says, no contradictions have arisen from this concept:

[it] has never led to any antinomy whatsoever: that is, the perfectly "naive" and uncritical working with this concept of set has so far proved completely self-consistent. (Gödel 1990, p. 259)

Favoring the iterative concept of set does not mean that mathematicians really (should) think that all objects in mathematics are iterative sets. The representation of everything as iterative sets is just a way to find a common ground on which all of mathematics can be understood.

3.2   The multiverse

There are at least two sources of possible variation in the cumulative hierarchy. The first is the power-set operation and the second is the length of the iteration. The latter is less interesting from the multiverse point of view. If we adopt a new Axiom of Infinity, such as the existence of a strongly inaccessible cardinal κ, we have immediately two universes, V_κ and V. However, we should not think of them as "parallel" universes. In fact, V_κ is rather an initial segment of V and exists as a set in V. Moreover, truth in V_κ is definable in V. There is no reason to think of V_κ as a parallel universe to V, one which we cannot distinguish from V and one about which we do not know whether it is the same as V or not. It is rather the opposite. We know that $V \neq V_\kappa$ since V_κ is a set and V_κ satisfies "there are no inaccessible cardinals" (if we chose κ to be the first inaccessible cardinal), unlike V. The fact that ZFC does not decide (if the existence of inaccessible cardinals is consistent) the truth of "there are no inaccessible cardinals" does not mean that we cannot assign a truth-value to this statement. We would simply say that the statement is false because its negation is a new axiom that we have adopted.

The other possible source of variability in the cumulative hierarchy is the power-set operation. Lindström (2000) presents a detailed analysis of the problem of the power-set operation. He accepts the power-set of $\mathbb{N}$ because he can *visualize* $\mathcal{P}(\mathbb{N})$ as the sets of infinite branches of the full binary tree $2^{<\omega}$. But, says Lindström, we have no way of visualizing the power-set of $\mathcal{P}(\omega)$, i.e., the set $\mathcal{P}\,(\mathcal{P}\,(\omega))$. In other words, Lindström sees no problem with the infinite sets $\mathbb{N}$ and $\mathbb{R}$ but the set of all sets of reals he finds problematic, because he does not know how to form a picture of it in his mind. As it happens, it is exactly the set of sets of reals that also decides the problematic CH.[15]

Lindström emphasizes the role of visualization in making sense of set theory, and finishes by criticizing the power-set operation. A different view of set theory acknowledges that there are unvisualizable sets, such as a well-ordering of the reals, but these "random" sets are necessary for a smooth development of set theory and therefore they are accepted. So sets are roughly divided into two categories: the "simple" sets, studied in descriptive set theory, and the "arbitrary" sets, studied in the more abstract areas of set theory.

It is in problems related to the power-set operation, and more generally to the "arbitrary" sets, where the multiverse idea emerges. It may just be the nature of the power-set operation that it eludes uniqueness. In anticipation of this we leave the uniqueness untouched and allow different cumulative hierarchies to emerge, in "parallel." Note that this does not mean that we abandon the Power-Set Axiom, which only says that whatever subsets of a given set we happen to have, they can be collected together.

If we allow different power-sets to emerge, we should not be able to talk about them explicitly. For example, if we had two power-sets $\mathcal{P}_1(X)$ and $\mathcal{P}_2(X)$ for a set X, we could immediately use the Axiom of Extensionality (which we will assume) to derive $\mathcal{P}_1(X) = \mathcal{P}_2(X)$. So the universes are completely hidden from each other.

We cannot name individual universes by any means. Some parallel universes may be distinguishable by their properties. For example, if there are parallel universes satisfying $V = L$, we can use the sentence $V \neq L \vee \varphi$ to say that φ is true in those universes. However, we cannot name individual universes. If we could, we would be able to say what is in that universe and what is not, while at the same time the universes should be "everything." What is the use of multiverse set theory if we cannot say anything about the individual universes? The point of the multiverse is

---

[15] If ZFC is consistent, there are two models of ZFC which have the same ordinals, cardinals, and reals, but which differ on CH (folklore, personal communication from Matt Foreman).

that it makes it possible to think of the set-theoretic reality as a definite well-defined structure and still doubt the uniqueness of the power-set operation, and keep open the possibility of absolute undecidability. Absolute undecidability has in the multiverse then a rational explanation, rather than being left as a sign of vague uncertainty. In Section 5 we extend our framework to allow more access to the internal structure of the multiverse without being able to name the universes.

Let us finally build the cumulative multiverse. It looks very familiar:

$$\begin{cases} \mathbb{V}_0 & = \emptyset \\ \mathbb{V}_{\alpha+1} & = \mathcal{P}(\mathbb{V}_\alpha) \\ \mathbb{V}_\nu & = \cup_{\alpha<\nu}\mathbb{V}_\alpha, \text{ if } \nu \text{ limit} \\ \mathbb{V} & = \cup_\alpha\mathbb{V}_\alpha. \end{cases} \tag{1}$$

The ordinals used in the equations (1) are ordinals simultaneously in all the parallel universes. Also the empty set $\emptyset$ is the empty set in each universe separately. This seems unnecessary because surely there is no variability or randomness in the empty set. Indeed, so far we have not introduced a way of seeing whether the empty sets of different universes are all equal or not. Intuitively there is every reason to believe that they are equal. When we come to higher and higher levels V_α there is less and less reason to believe that the versions of V_α in the different universes are the same. In Section 5 we will address this issue in detail.

3.3 Axioms

The axioms of the multiverse are the usual axioms of ZFC written in the vocabulary $\{\in\}$ using first-order logic, with variables ranging over sets in the multiverse, as if there were only one universe. In a sense, the variables are thought to range simultaneously over all universes This means that we do not have any symbols for relations such as "being in the same universe," no names for individual universes, etc. Moreover, we write the axioms in ordinary first-order logic. There are no new logical operations arising from the multiverse perspective at this point. In Section 5 we consider the use of extensions of first-order logic.

The meaning of the *Axiom of Extensionality* $\forall x \forall y \ (\forall z \ (z \in x \leftrightarrow z \in y) \rightarrow x = y)$ is that two sets are equal if they have the same elements. Since our mental picture of the universe is that it consists of parallel realities, and we think of bound variables as ranging over each universe at the same time, as if the multiverse consisted of a Cartesian product of universes, each set that we consider in the axioms has its "versions" (or "projections," although there is no such projection map in the language of set theory) in the different universes. Thus the Axiom of Extensionality says that two sets from the multiverse are equal if they have the same elements in each of those parallel universes. Note that the empty sets of the different universes do not have elements but they are not *urelements*, for the Axiom of Extensionality implies that no universe has urelements (apart from the empty set). The meaning of the *Axiom of Pairing* $\forall x \forall y \exists z \forall u\ (u \in z \leftrightarrow (u = x \lor u = y))$ is that from any two sets a and b we can form the unordered pair in each universe separately and the result is denoted $\{a, b\}$. As to the *Axiom of Union* $\forall x \exists y \forall z (z \in y \leftrightarrow \exists u (u \in x \land z \in u))$, the set $\cup\ a$ has its version in each universe, always the union of the sets in that universe that are elements of the corresponding version of a. Concerning the *Axiom of Power-Set* $\forall x \exists y \forall z\ (z \in y \leftrightarrow \forall u\ (u \in z\ u \in x))$ as discussed above, even if the set a is the same in each universe, the power-set is allowed to be different, although the power-sets may also all be equal, as far as the ZFC axioms can tell.

In the *Axiom Schema of Subsets*

$$\forall x \forall u_1 \ldots \forall u_n \exists y \forall z (z \in y \leftrightarrow (z \in x \land \varphi(z, \vec{u})))$$

we have a formula $\varphi(z, \vec{y})$ that we use to cut out a subset from a given set x. The quantifiers of the first-order formula $\varphi(z, \vec{u})$ range over the multiverse elements of x. Notice that the same definition is used in each universe to cut out a subset from x. It is the same with the *Axiom Schema of Replacement*

$$\forall x \forall w_1 \ldots \forall w_n (\forall u \forall z \forall z'\ ((u \in x \land \varphi(u, z, \vec{w}) \land \varphi(u, z', \vec{z})) \rightarrow z = z')$$
$$\rightarrow \exists y \forall z (z \in y \leftrightarrow \exists u (u \in x \land \varphi(u, z, \vec{w})))).$$

The *Axiom of Infinity* $\exists x (\exists y (y \in x \land \forall z \neg (z \in y)) \land \forall y (y \in x \rightarrow \exists z (y \in z \land z \in x)))$ says that every universe of the multiverse has an infinite set. This has, of course, nothing to do with the question whether there are infinitely many universes. We have no axiom (yet) which would imply that. The *Axiom of Foundation* $\forall x \exists y (x \cap y = \emptyset)$ arises in multiverse set theory for the same reason as in one universe set theory: sets are formed in stages and the stages are well founded, so every set x has an element y which was formed earliest. The "choice" involved in the *Axiom of Choice* is in multiverse set theory spread over the multiverse. The ideology of multiverse set theory is that "choosing" happens simultaneously in all universes. It is exactly because the "choosing" is problematic even in one universe when there are infinitely many sets to choose from, that we allow many universes, each with its own mode of choosing. The extra complication arising from choosing simultaneously in many universes is simply not part of the setup of multiverse set theory.

4 Multiverse logic

Let us now discuss the metamathematics of multiverse set theory, that is, we take the (codes of the) ZFC axioms as a set inside set theory, define what it means for another set to be a multiverse model of the set of ZFC axioms, again inside set theory, and then investigate in set theory this relationship. It makes no difference whether we study the metamathematics of ZFC in one universe set theory or multiverse set theory, because we identify truth in both cases with truth in all (or in *the one*, if there is only one) universes.

Instead of the concept of a single model with one universe satisfying ZFC, we have the concept of a *multiverse model* satisfying ZFC.

Let us first define the general concept of a multiverse model in first-order logic:

Definition 4.1 *Suppose L is a vocabulary. A multiverse L-model is a set* $\mathbb{M}$ *of L-structures.*

So our concept of a multiverse model has the same degree of generality as the concept of a model in first-order logic. There need not be any connections between the individual models constituting a multiverse model.

Example 4.2 *The following are examples of L-multiverse structures.*

1. *The **empty** multiverse. This is a singular case but permitted by the definition.*
2. *The **one universe** multiverse* $\{M\}$ *for any L-structure M. This is the classical one universe structure.*
3. *The **full** multiverse: suppose T is a countable first-order L-theory. The set of all models of T, with domain* $\mathbb{N}$, *is an L-multiverse. This is* mutatis mutandis *the set of complete consistent extensions of T.*
4. *A **bifurcated** multiverse: suppose T is a countable first-order L-theory and* φ *is an L-sentence such that* $T \nvdash \varphi$ *and* $T \nvdash \neg\varphi$. *Let* $M_0 \models T \cup \{\varphi\}$ *and* $M_1 \models T \cup \{\neg\varphi\}$. *Then* $\mathbb{M} = \{M_0, M_1\}$ *is a multiverse L-model of T. When we define truth in a multiverse (see Definition 4.1), we shall see that T holds in* $\mathbb{M}$ *but neither* φ *nor* $\neg\varphi$ *does.*

5. *Woodin's **generic** multiverse of set theory* (Woodin 2011): *suppose M is a countable transitive model of ZFC. Let $\mathbb{M}$ be the smallest set of countable transitive sets such that $M \in \mathbb{M}$ and such that for all pairs, (M_1, M_2), of countable transitive models of ZFC such that M_2 is a generic extension of M_1, if either $M_1 \in \mathbb{M}$ or $M_2 \in \mathbb{M}$ then both M_1 and M_2 are in $\mathbb{M}$. It is a remarkable property of this multiverse model that if we start to build it replacing M by any $N \in \mathbb{M}$, the same $\mathbb{M}$ results.*

6. *Steel's **generic** multiverse of set theory* (see Chapter 8). *Let M be a transitive model of ZFC, and let G be M-generic for* $\mathrm{Col}(\omega, < \mathrm{OR}^M)$. *The worlds of the multiverse* $\mathbb{M}^G$ *are all those W such that* $W[H] = M[G \restriction \alpha]$, *for some H set generic over W, and some* $\alpha \in \mathrm{OR}^M$. *Again, if we start to build this replacing M by any $N \in \mathbb{M}^G$, the same $\mathbb{M}^G$ results.*

7. *The set of countable computably saturated models of ZFC, with domain $\mathbb{N}$, is a multiverse of set theory introduced in* Gitman and Hamkins (2010), *where it is shown to satisfy the hypotheses for multiverse models of set theory of* Hamkins (2011).

4.1   Metamathematics

We proceed now towards the truth definition in multiverse logic. We work inside multiverse set theory, treating formulas and multiverse models as sets on a par with other sets. So a multiverse model $\mathbb{M}$ exists intuitively at the same time in possibly more than one universe. However, we need not worry about this because the metamathematical study of the multiverse uses only first-order logic, that is, we do not use here the new logical operations of Section 5, and everything we say is consistent with there being just one universe. So we can avoid the infinite regress of multiverse, multimultiverse,...etc.

Since a multiverse model is just a set of models, the set-theoretic operations make sense. The *union* of two multiverse L-models $\mathbb{M}$ and $\mathbb{M}'$ is their set-theoretic union, denoted $\mathbb{M} \cup \mathbb{M}'$.

Definition 4.3 *Suppose $\mathbb{M}$ is an L-multiverse model and $M \in \mathbb{M}$. An assignment into M is a mapping s such that $\mathrm{dom}(s)$ is a set of variables and $s(x) \in M$ for each $x \in \mathrm{dom}(s)$. An assignment in $\mathbb{M}$ is a mapping such that $\mathrm{dom}(\mathfrak{s}) = \mathbb{M}$ and if $M \in \mathbb{M}$ then $\mathfrak{s}(M)$, denoted $\mathfrak{s}_M$, is an assignment into M.*

We write $M \models_s \varphi$ if, according to the usual textbook definition, the assignment s satisfies the first-order formula φ in the one universe model M.

As is the custom with multiverses (e.g., Woodin 2011), the truth in a multiverse simply means truth in each structure of the multiverse:

Definition 4.4 *Suppose* $\mathbb{M}$ *is an L-multiverse structure,* $\mathfrak{s}$ *an assignment in* $\mathbb{M}$, *and* φ *a first-order L-formula. We define*

$$\mathbb{M} \models_{\mathfrak{s}} \varphi \Leftrightarrow_{def} \forall M \in \mathbb{M}\,(M \models_{\mathfrak{s}(M)} \varphi). \tag{2}$$

For sentences φ *we ignore* $\mathfrak{s}$.

An immediate consequence of the definition is that validity in all multiverse models is equivalent to validity in all one universe models.

If we consider truth in a multiverse as an *independent* concept, we can note the following consequence of the definition. $\mathfrak{s}\!\restriction\!\mathbb{M}_0$ means the restriction of the mapping $\mathfrak{s}$ to the set $\mathbb{M}_0$. For the interpretation of the quantifiers we adopt the following notation. If F is a mapping such that $\mathrm{dom}(F) = \mathbb{M}$ and $F(M) \in M$ for each $M \in \mathbb{M}$, then $\mathfrak{s}(F/x)$ denotes the modification of s such that $\mathfrak{s}(F/X)(M)(x) = F(M)$ and $\mathfrak{s}(F/x)(M)(y) = \mathfrak{s}(M)(y)$ for $y \neq x$. The value of a term t in a (one universe) model M under the assignment s is denoted by $t^M\langle s\rangle$.

Proposition 4.5 *Truth in a multiverse has the following properties:*

1. $\mathbb{M} \models_{\mathfrak{s}} t = t'$ *if and only if* $t^M\langle \mathfrak{s}_M\rangle = t'^M\langle \mathfrak{s}_M\rangle$ *for all* $M \in \mathbb{M}$.
2. $\mathbb{M} \models_{\mathfrak{s}} \neg t = t'$ *if and only if* $t^M\langle \mathfrak{s}_M\rangle \neq t'^M\langle \mathfrak{s}_M\rangle$ *for all* $M \in \mathbb{M}$.
3. $\mathbb{M} \models_{\mathfrak{s}} R(t_1, \ldots, t_n)$ *if and only if* $(t_1^M\langle \mathfrak{s}_M\rangle, \ldots, t_n^M\langle \mathfrak{s}_M\rangle) \in R^M$ *for all* $M \in \mathbb{M}$.
4. $\mathbb{M} \models_{\mathfrak{s}} \neg R(t_1, \ldots, t_n)$ *if and only if* $(t_1^M\langle \mathfrak{s}_M\rangle, \ldots, t_n^M\langle \mathfrak{s}_M\rangle) \notin R^M$ *for all* $M \in \mathbb{M}$.
5. $\mathbb{M} \models_{\mathfrak{s}} \varphi \wedge \psi$ *if and only if* $(\mathbb{M} \models_s \varphi$ *and* $\mathbb{M} \models_{\mathfrak{s}} \psi)$.
6. $\mathbb{M} \models_{\mathfrak{s}} \varphi \wedge \psi$ *if and only if there are* $\mathbb{M}_0 \subseteq \mathbb{M}$ *and* $\mathbb{M}_1 \subseteq \mathbb{M}$ *such that* $\mathbb{M} = \mathbb{M}_0 \cup \mathbb{M}_1$, $\mathbb{M}_0 \models_{\mathfrak{s}} \wedge \varphi$ *and* $\mathbb{M}_1 \models_{\mathfrak{s}} \wedge \psi$.
7. $\mathbb{M} \models_{\mathfrak{s}} \exists x \varphi$ *if and only if there is* F *such that* $\mathrm{dom}(F) = \mathbb{M}$, $F(M) \in M$ *for all* $M \in_{\mathfrak{s}} \mathbb{M}$, *and* $\mathbb{M} \models_{\mathfrak{s}(F/x)} \varphi$.
8. $\mathbb{M} \models_{\mathfrak{s}} \forall x \varphi$ *if and only if for all* F *such that* $\mathrm{dom}(F) = \mathbb{M}$ *and* $F(M) \in M$ *for all* $M \in \mathbb{M}$, *we have* $\mathbb{M} \models_{\mathfrak{s}(F/x)} \varphi$.

We could have used the above conditions, instead of Definition 4.4, to define truth. The "value" of a term t in a multiverse $\mathbb{M}$ under the assignment $\mathfrak{s}$ can be thought of as the mapping $M \mapsto t^M\langle \mathfrak{s}_M\rangle$, where $M \in \mathbb{M}$.

The intuition behind the property (6) above is the following. For the disjunction $\varphi \vee \psi$ to hold in the multiverse, every model $M \in \mathbb{M}$ has to decide of $\{\varphi, \psi\}$, which is true in M. We collect into $\mathbb{M}_0$ those M which pick φ, and into $\mathbb{M}_1$ those that pick ψ. Since conceivably no M picks, say, ψ, we may end up with $\mathbb{M}_1 = \varnothing$. This is the reason for allowing the empty multiverse.

For the negation $\mathbb{M} \models_s \neg\varphi$, the usual $\mathbb{M} \not\models_s \varphi$ will not work, since it is possible in view of Definition 4.4 that

$$\mathbb{M} \not\models_s \varphi \text{ and } M \not\models_s \neg\varphi.$$

For example, for the generic multiverse of Example 4.2 (5) we have

$$\mathbb{M} \not\models CH \text{ and } \mathbb{M} \not\models \neg CH.$$

Negated formulas are therefore handled by pushing the negation into the formula by means of the de Morgan laws.[16]

As validity in all multiverse models is equivalent to validity in all one universe models, a first-order sentence is valid in all multiverse models if and only if it has a proof in the ordinary sense of first-order logic. Until we extend first-order logic with new logical operations (Section 5) it is only metamathematics and intuition that distinguishes multiverse logic from one universe logic. Proving things is the same in each case. The different intuition is not manifested in proofs in any way.

After this introduction to multiverse logic let us return to the meta-mathematics of multiverse set theory.

We have the first-order axioms ZFC and the concept of a multiverse model of them. The generic multiverses are certainly very good examples of multiverse models of ZFC. As was mentioned, a particularly attractive feature is their invariance under permutation of the inital model. So if one works inside one of the universes M of a generic multiverse $\mathbb{M}$ one can really forget about the particular M, be oblivious of which of the many universes in $\mathbb{M}$ this M happens to be. Moreover, in the above two generic multiverses one can also express the truth of any given φ in all the universes of the multiverse with a translation φ^*. Thus

$$M \models \varphi^* \Leftrightarrow \mathbb{M} \models \varphi.$$

Steel (in Chapter 8) presents a formal system MV for multiverse set theory. He has two sorts for variables, set sort and world sort, and axioms which dictate that any multiverse model of MV is of the form of a generic multiverse. The difference from our multiverse set theory is first of all that

---

[16] $\neg(\varphi \wedge \psi) \equiv \neg\varphi \vee \neg\psi$, $\neg\exists x\varphi \equiv \forall x\neg\varphi$, etc.

we have no variables for worlds, as the worlds in our system are completely hidden, and second that Steel has axioms which force the multiverse to be a generic multiverse, while at least for the time being we allow any kind of multiverse, even empty or full. We will say more of the connections to generic multiverse in Section 5.3.

4.2   *Truth in the multiverse*

Having introduced the axioms of multiverse set theory and having taken the first steps in the study of the metamathematics of multiverse ZFC, we return to the basic questions of truth and justification in the multiverse framework.

Propositions of multiverse set theory are of the form

$$\Phi(a_1, \ldots, a_n), \tag{3}$$

where $\Phi(x_1, \ldots, x_n)$ is a first-order formula[17] and $a_1, \ldots, a_n$ are definable[18] terms. The meaning of (3) is intuitively that $\Phi(a_1, \ldots, a_n)$ is true in the multiverse $\mathbb{V}$, which for first-order formulas means intuitively simply truth in each universe separately. The bound variables of $\Phi(x_1, \ldots, x_n)$ range intuitively over all the universes of $\mathbb{V}$ simultaneously. We use the ordinary rules of classical logic to derive truths from the axioms. Even though we do not accept the principle that nontruth of $\neg\varphi$ implies φ, all the rules of logic are valid. In particular, $\neg\neg\varphi$ logically implies φ.

The proposition (3) has definable terms $a_1, \ldots, a_n$ and in the spirit of the multiverse framework we think of them as existing simultaneously but independently in each universe. First-order logic does not provide the means to force such terms to be equal across the universes. Later in Section 5 we introduce tools to overcome this.

Let us then consider the question under what circumstances are we justified in asserting (3). The gold standard of justification in set theory – also multiverse set theory – is a proof of (3) from the ZFC axioms. A widely accepted addition to ZFC are large cardinal axioms in one form or another. This justification, using large cardinals or not, gives more than is needed, for if we subject ZFC to a metamathematical study, a proof in ZFC justifies truth in *all* multiverses which satisfy the axioms. However, from

---

[17] In Section 5 we allow propositions that arise naturally in the multiverse framework but go beyond first-order logic.

[18] The concept of a definable term is not itself definable but for each quantifier-rank separately this concept is definable.

the point of view of multiverse set theory as a foundation of mathematics there is just one "true" multiverse.

By the laws of classical logic, still valid in the multiverse framework, we accept $\varphi \vee \neg\varphi$ as true for every φ. How is this justified? In terms of justification by proof the matter is not resolved because $\varphi \vee \neg\varphi$ is either an axiom or follows from essentially equivalent axioms with a similar need for justification. The justification of the Law of Excluded Middle $\varphi \vee \neg\varphi$ for first-order φ reduces to its justification, which we assume, in the one universe framework, for the meaning of $\varphi \vee \neg\varphi$ is that every single universe satisfies φ or $\neg\varphi$.

By Gödel's incompleteness theorem Con(ZFC) cannot be decided on the basis of ZFC alone, unless ZFC is inconsistent. From the perspective of the *full* multiverse, Con(ZFC) is absolutely undecidable. But this is no basis to consider Con(ZFC) absolutely undecidable in $\mathbb{V}$. In fact, the large cardinal axioms decide it. Still, the method by which incompleteness arises in Gödel's incompleteness theorem manifests a kind of absolute undecidability, not of a particular proposition, but of the informal expression "the method of justification is itself consistent."

5 Multiverse and team semantics

The semantics of multiverse logic suggests new logical operations with applications to multiverse set theory.

The basic idea is the following. Even if we think that the universe of set theory may be bifurcated, we need *not* think that every possible complete extension of ZFC is realized in the multiverse. It is very plausible to think that the natural numbers and even the real numbers are the same in all universes of the multiverse. (In the one universe set theory the whole idea that the natural (or the real) numbers are the same is of course meaningless.)

We may also have doubts about the truth-value of CH but we may be convinced that whether CH holds is independent of whether there are inaccessible cardinals. Again, in the one universe set theory it is meaningless to ask whether CH is independent of the existence of inaccessible cardinals. In one universe set theory CH is true or false and there are inaccessible cardinals or there are none, but from these facts one cannot derive conclusions about mutual independence. The fact that there are models of ZFC which demonstrate the independence of ZFC from large cardinals is of no help because those models are just *some* models, not necessarily models in *our* multiverse.

There are several ways one can "homogenize" the multiverse, distinguish it from the trivializing full multiverse of all possible models of ZFC, and bring it closer to the world of one universe. One approach is to have variables for worlds, as Steel (Chapter 8) does, and specify axioms which dictate that all the worlds are related to each other by forcing. We choose another route, that of extending the logical power of first-order logic by new logical operations, based on *team semantics*.

Team semantics[19] is a variation of ordinary Tarski semantics of first-order logic. In team semantics the basic concept is not that of an assignment *s* satisfying a formula φ in a model *M*, but a set *X* of assignments, called *a team*, satisfying a formula. For first-order formulas a team satisfies the formula simply if all the assignments satisfy the formula. What is gained by the introduction of teams? In a team one can manifest dependence and independence phenomena. In the extension of first-order logic called *dependence logic* (Väänänen 2007) the following new atomic formulas are added to first-order logic:

$$=(\vec{y}, \vec{x}),$$

with the so-called Armstrong Axioms (Armstrong 1974), governing the intuition "the values of $\vec{y}$ functionally determine the values of $\vec{x}$ in the team." Semantically, the truth of $=(\vec{y}, \vec{x})$ in a team *X* is defined as

$$\forall s, s' \in X \left(s(\vec{y}) = s'(\vec{y}) \rightarrow s(\vec{x}) = s'(\vec{x}) \right).$$

For example, the sentence

$$\forall z \forall x \exists y \left(=(y, x) \land \neg y = z \right)$$

says that there is a one-one function from the universe to a proper subset, i.e., the universe is infinite. According to Henkin (1961), the idea goes back to A. Ehrenfeucht. The idea is the following. In an infinite universe we can pick a one-one function *f* into the complement of the element *z*. Then we let $y = f(x)$. Because *f* is one-one, the value of *y* completely determines the value of *x*. Conversely, if for given *x* we can find $y \neq z$ such that *x* is a function $g(y)$ of *y*, then the mapping $x \mapsto y$ must be one-one, and hence the domain must be infinite. Thus validity in dependence logic is nonaxiomatizable. However, the first-order logical consequences of dependence logic sentences can be axiomatized and such axioms are given in Kontinen and Väänänen (2013).

---

[19] See Väänänen and Hodges (2010) for the origin of team semantics and Väänänen (2007) for a detailed study of it.

There are many other new atomic formulas that suggest themselves in the team semantics context, for example the *independence atom* (Grädel and Väänänen 2013) $x \perp y$ with the meaning that x and y are "independent" (the values of x reveal nothing about the values of y), and the *inclusion atom* (Galliani 2012) $x \subseteq y$ with the meaning that values of x occur also as values of y,

The connection between multiverse logic and dependence logic is that the same feature of team semantics which allows us to express the dependence and independence of variables, allows us in the multiverse framework to talk about dependence and independence of formulas, and about absolute undecidability. We accomplish this by thinking of a team of assignments as a multiverse of copies of a single model with single assignments. We now make that shift.

Definition 5.1 *Multiverse dependence logic, MD for short, is the extension of first-order logic by the dependence atoms* $=(\vec{y}, \vec{x})$. *(Later we add other new logical operations.)*

Inference in MD takes place as in first-order logic, using special axioms and rules for $=(\vec{y}, \vec{x})$, given in Kontinen and Väänänen (2013).

The semantics of MD in the multiverse setup is as follows. To account for the fact that there may be many identical copies of the same model in the multiverse, we assume that multiverses are indexed $\mathbb{M} = \{M_i : i \in I\}$. The semantics is clearly independent of what indexing is used. Assignments $\mathfrak{s}$ are defined on I: $\mathfrak{s}_i$ is the assignment of the model M_i.

Adapting the conditions of Definition 4.3 to the team semantics yields the following.[20]

Definition 5.2 *Suppose* $\mathbb{M} = \{M_i : i \in I\}$ *is an L-multiverse structure,* $\mathfrak{s}$ *a team in* $\mathbb{M}$, *and* φ *a dependence logic L-formula. We define* $\mathbb{M} \models_{\mathfrak{s}} \varphi$ *as follows.*

(i) $\mathbb{M} \models_{\mathfrak{s}} t \doteq t'$ *if and only if* $t^M \langle \mathfrak{s}_i \rangle = t'^M \langle s_i \rangle$ *for all* $i \in I$.

(ii) $\mathbb{M} \models_{\mathfrak{s}} \neg t = t'$ *if and only if* $t^M \langle \mathfrak{s}_i \rangle \neq t'^{Mi} \langle s_i \rangle$ *for all* $i \in I$.

(iii) $\mathbb{M} \models_{\mathfrak{s}} R(t_1, \ldots, t_n)$ *if and only if* $(t_1^{M_i} \langle \mathfrak{s}_i \rangle, \ldots, t_1^{M_i} \langle \mathfrak{s}_i \rangle) \in R^{M_i}$ *for all* $i \in I$.

---

[20] It makes again no difference whether we investigate the semantics of dependence logic in one universe set theory or multiverse set theory. The reason is that, although dependence logic goes beyond first-order logic and dependence logic truth cannot be reduced to truth in all universes, we do *not* use dependence logic in metatheory. Our metatheory is just first-order logic.

(iv) $\mathbb{M}\models_{\mathbf{s}}\neg R(t_1, \ldots, t_n)$ *if and only if* $(t_1^{M_i}\langle \mathbf{s}_i\rangle, \ldots, t_1^{M_i}\langle \mathbf{s}_i\rangle) \notin R^{M_i}$ *for all* $i \in I$.

(v) $\mathbb{M}\models_{\mathbf{s}} = (\vec{y}, \vec{x})$ *if and only if* $\forall i, i' \in I(\mathbf{s}_i(\vec{y}) = \mathbf{s}_{i'}(\vec{y}) = \mathbf{s}_{i'}(\vec{y}) \rightarrow \mathbf{s}_i(\vec{x}) = \mathbf{s}_{i'}(\vec{x}))$.

(vi) $\mathbb{M}\models_{\mathbf{s}}\varphi \wedge \psi$ *if and only if* $(\mathbb{M}\models_{\mathbf{s}}\varphi \text{ and } \mathbb{M}\models_{\mathbf{s}}\psi)$.

(vii) $\mathbb{M}\models_{\mathbf{s}}\varphi \vee \psi$ *if and only if there are* $\mathbb{M}_0 \subseteq \mathbb{M}$ *and* $\mathbb{M}_1 \subseteq \mathbb{M}$ *such that* $\mathbb{M} = \mathbb{M}_0 \cup \mathbb{M}_1$, $\mathbb{M}_0 \models_{\mathbf{s}\restriction\mathbb{M}_0}\varphi$ *and* $\mathbb{M}_1 \models_{\mathbf{s} \restriction \mathbb{M}_1}\psi$.

(viii) $\mathbb{M}\models_{\mathbf{s}}\exists x\varphi$ *if and only if there is* F *such that* $\mathrm{dom}(F) = I$, $F(i) \in M_i$ *for all* $i \in I$, *and* $\mathbb{M}\models_{\mathbf{s}'}\varphi$, *where for each* $i \in I$, $\mathbf{s}'_i$ *is the assignment* $\{\mathbf{s}_i(F(i)/x):i \in I\}$.

(ix) $\mathbb{M}\models_{\mathbf{s}}\forall x\varphi$ *if and only if* $\mathbb{M}'\models_{\mathbf{s}'}\varphi$, *where* $\mathbb{M}' = \{M'_i : i \in I'\}$, $I' = \{(M, a) : M \in \mathbb{M}, a \in M\}$, $M'_i = M, \mathbf{s}'_i = \mathbf{s}_i(a/x)$, *for* $i = (M, a)$.

For example, $\mathbb{M} \models_{\mathbf{s}} =(x)$ if and only if $\forall M, M' \in \mathbb{M}$ $(\mathbf{s}_M(x) = \mathbf{s}_{M'}(x))$, that is, if and only if x has a constant value over all models. Thus $\mathbb{M} \models \exists x =(x)$ means that all models have a common element.

Compared to the conditions in Proposition 4.5, there is a difference. Condition (8) of Proposition 4.5 has become condition (ix) of Definition 5.2. The intuition behind the new condition (ix) is that the team $\mathbf{s}'$ presents all the possible values of x at once, in one big multiverse $\mathbb{M}'$, where every possible value for x occurs in some universe, while the old (8) goes through different choices $F(M)$ for x, one function at a time. So the new (ix) has all the functions F at once in a big Cartesian product, while the old (8) checked each function F separately. From the point of view of first-order truth there is no difference whether one checks these one at a time or all at once. But from the point of view of dependence logic it is necessary to use the new all-at-once definition (ix). Another change to Proposition 4.5 is the inclusion of the case (v) for the dependence atom.

5.1 *Homogenization*

We now introduce new logical operations which are multiverse specific. The goal is to homogenize the multiverse, because the intuition about the multiverse is not that everything that is logically possible should also happen in some universe (which would lead to the *full* multiverse), but that the multiverse is one universe the boundaries of which are "*verschwommen*" ("blurred"), as von Neumann wrote (see above).

The first new logical operation is the following.

Definition 5.3 (*Boolean disjunction*) $\mathbb{M} \models_s \varphi \vee_B \psi$ *if and only if* $\mathbb{M} \models_s$ φ *or* $\mathbb{M} \models_s \psi$.

This may look a bit surprising – do we not already have disjunction in first-order logic? However, our prior definition of the semantics of disjunction was different: $M \models_s \varphi \vee \psi$, which in the case of first-order logic means $\forall M \in \mathbb{M}(M \models_s \varphi \text{ or } M \models_s \psi)$, certainly different from "$\forall M \in \mathbb{M}(M \models_s \varphi)$ *or* $\forall M \in \mathbb{M}(M \models_s \psi)$."

Let us write

$$=(\varphi) \text{ for } \varphi \vee_B \neg\varphi.$$

This says that φ has a truth-value, i.e., it is either true in the entire multiverse or false in the entire multiverse. If φ says "$o^\#$ exists," then the generic multiverse satisfies $=(\varphi)$. Note that $=(\varphi)\vee =(\varphi)$ and $=(\varphi)$ are in general nonequivalent. The first says that the multiverse can be divided into two parts in both of which φ has a (possibly different) truth-value, while $=(\varphi)$ says φ has the same truth-value in the entire multiverse. For example, if φ says "$o^\#$ exists," and the multiverse is the union of two generic universes, one with φ and another with $\neg\varphi$, then the multiverse satisfies $=(\varphi)\vee =(\varphi)$ but not φ, and also not $=(\varphi)\vee_B =(\varphi)$.

In terms of logical rules the difference between disjunction and Boolean disjunction is:

$$
\begin{array}{ccc}
[\varphi] \quad [\psi] & & [\varphi] \quad [\psi] \\
\vdots \quad \vdots & & \vdots \quad \vdots \\
\dfrac{\varphi \vee \psi \quad \theta \quad \theta}{\theta \vee \theta} & & \dfrac{\varphi \vee_B \psi \quad \theta \quad \theta}{\theta}
\end{array}
$$

We have allowed, mainly for technical reasons, the possibility that the multiverse is empty. Now we introduce a logical constant NE to remedy this.

Definition 5.4 (*Nonemptiness*) $\mathbb{M} \models_s \text{NE} \Leftrightarrow \mathbb{M} \neq \varnothing$.

This seems to be a most unnecessary addition, because we could have assumed all along that multiverses are nonempty. However, since we did not make this assumption, we have to introduce NE. There is no first-order formula which would be able to say that the multiverse is nonempty. For example, the empty multiverse satisfies $\varphi \wedge \neg\varphi$ for all φ.

With NE we can introduce a new logical operation. We write

$$\neq(\varphi) \text{ for } (\varphi \wedge \text{NE}) \vee (\neg\varphi \wedge \text{NE}).$$

The meaning of $\neq(\varphi)$ is that φ is *absolutely undecidable,* true in some models and false in some. So if we add $\neq$(CH) to the axioms of set theory we are committed to the idea that the continuum hypothesis will never be solved.

What is the evidence we can give to $\neq$(CH)? If we adopted it as an axiom, we should think that it is self-evident. The only thing that is evident, however, is that despite over 130 years of attempts, there is no unanimous opinion among experts about it. Note that $\neq$(CH) cannot have first-order consequences which would not already follow from ZFC.

On the other hand, adding =(CH) to the axioms of set theory would indicate conviction that CH has a truth-value, although we do not know, and may never know, what the truth-value is.

We can also say that the universes have the same arithmetic theory of natural numbers by stipulating $=(\varphi^N)$ for every arithmetical sentence φ.

Using the Boolean disjunction and NE we can also define for first-order φ and ψ

$$\varphi \perp \psi$$

with the intuitive meaning that φ and ψ are *independent* in the multiverse. More exactly,

$$\mathbb{M} \models_s \varphi \perp \psi \Leftrightarrow$$

$$\forall M, M' \in \mathbb{M} \exists M'' \in \mathbb{M}[(M'' \models_{s_{M''}} \varphi \Leftrightarrow M \models_{s_M} \varphi)$$

$$\text{and } (M'' \models_{s_{M''}} \psi \Leftrightarrow M' \models_{s_{M'}} \psi).]$$

This is reminiscent of the formula $\vec{x} \perp \vec{y}$ in Grädel and Väänänen (2013) (see also Galliani 2012). The idea is that M'' picks the truth-value of φ from M and the truth-value of ψ from M'. Thus, in the light of the multiverse $\mathbb{M}$, knowing the truth-value of φ gives no clue as to the truth-value of ψ. It is still possible that $\mathbb{M}$ is a one universe multiverse, but if there are several models then there have to be enough models to satisfy the independence.

Definition 5.5 *Suppose* $\mathbb{M} = \{M_i : i \in I\}$ *is an L-multiverse structure,* s *an assignment in* $\mathbb{M}$, *and* $\varphi(\vec{x})$ *a first-order L-formula. We define* $\mathbb{M} \models_s = (\vec{x} : \varphi(\vec{x}))$ *as follows:*

$$\mathbb{M} \models_s = (\vec{x} : \varphi(\vec{x})) \Leftrightarrow$$

$$\forall M, N \in \mathbb{M} : \{\vec{a} \in M^n : M \models \varphi(\vec{a})\} = \{\vec{a} \in N^n : N \models \varphi(\vec{a})\}.$$

For example, =(φ) is equivalent to =(:φ) for first-order sentences φ. Notice that =($x : x = x$) says that all the models of the multiverse have the same domain, and $=(\vec{x}: R(\vec{x}))$ says that they have the same interpretation of the relation R. Moreover, $=(x: \forall y \neg y \in x)$ says all models in the multiverse have the same $\varnothing$. We can continue like this writing axioms which say that all models in the multiverse have the same natural numbers, real numbers, etc.

We include all the above operations in MD, and now turn to the problem how to justify the truth of a sentence of MD in the multiverse. The meaning of a sentence of MD of the form $=(\vec{x}: \varphi(\vec{x}))$ can be given in terms of the different models constituting the multiverse *only* on the level of intuition, or as a mathematical property when the relationship $\mathbb{M} \models_s = (\vec{x}: \varphi(\vec{x}))$ is studied as a mathematical relation. For verification of a proposition of the form $=(\vec{x}: \varphi(\vec{x}))$ we need axioms and rules of proof.

5.2 Axiomatization

We have defined the new logical operations by giving their semantics. This should be complemented by the elimination and introduction rules for such operations. For details of a complete and explicit axiomatization of first-order consequence in dependence logic, we refer to Kontinen and Väänänen (2013).

In Section 4.2 we discussed the problem how to justify claims of truth in the multiverse of set theory, and we gave the ZFC axioms, perhaps extended by Axioms of Infinity, as a criterion. The same applies to the problem of justification in multiverse dependence logic. The rules of classical logic can be freely applied to first-order formulas but with dependence formulas more care is needed, as, for example $=(\varphi) \vee =(\varphi)$ and $=(\varphi)$ need not be equivalent. Reduction to first-order logic gives the following result.

Theorem 5.6 *The relation $T \models \varphi$ for recursive dependence logic theories T and first-order sentences φ is effectively axiomatizable in the vocabulary $\{\in\}$.*

Proof. We essentially reconstruct the multiverse in (one universe) first-order logic by means of new predicates. The details of this are straightforward, but since we also have new logical operations, we try to be as detailed as possible. Notice that the new predicates that we introduce are only for the sake of the proof, they are not part of the axiomatization.

Suppose U is a unary predicate symbol, W a binary predicate symbol and S an $(n + 1)$-ary predicate symbol. Intuitively, U is the set of universes, W codes the domain of each universe, and S is an assignment. Let $\Theta_n(U, W, S)$ be the first-order sentence

$$\forall x(U(x) \to \exists y\, W(x,y)) \land \forall u \forall x(W(u,x) \to U(x)) \land$$

$$\forall u \forall x_1 \ldots \forall x_n (S(u,x_1,\ldots,x_n) \to (U(u) \land \bigwedge_{i=1}^n W(u,x_i))) \land$$

$$\forall u \forall x_1 \ldots \forall x_{2n}((S(u,x_1,\ldots,x_n) \land S(u,x_{n+1},\ldots,x_{2n}) \to \bigwedge_{i=1}^n x_i = x_{n+i}) \land$$

$$\forall u(U(u) \to \exists x_1 \ldots \exists x_n S(u,x_1,\ldots,x_n)).$$

We associate with every dependence logic formula $\varphi(x_{i_1},\ldots,x_{i_n})$ first-order sentences $\tau_{1,\varphi}(U,\ W,\ S)$ and $\tau_{0,\varphi}(U,\ W,\ S)$, with the intended meaning that φ is true (respectively, false) in the multiverse coded by $U,\ W$ and S, as follows.

Case 1 Suppose $\varphi(x_{i_1},\ldots,x_{i_n})$ is an atomic formula. We let $\tau_{1,\varphi}$ $(U,\ W,\ S)$ be

$$\forall u \forall x_{i_1} \ldots \forall x_{i_n}\left(S(u,x_{i_1},\ldots,x_{i_n}) \to \varphi^*(u,x_{i_1},\ldots,x_{i_n})\right)$$

and we let $\tau_{0,\varphi}\,(U,\ W,\ S)$ be

$$\forall u \forall x_{i_1} \ldots \forall x_{i_n}\left(S(u,x_{i_1},\ldots,x_{i_n}) \to \neg\varphi^*(u,x_{i_1},\ldots,x_{i_n})\right)$$

where $\varphi^*(u,x_{i_1},\ldots,x_{i_n})$ is obtained from $\varphi(x_{i_1},\ldots,x_{i_n})$ by replacing every predicate symbol $R(t_1,\ldots,t_k)$ by $R^*(u,t_1,\ldots,t_k)$, every function symbol $F(t_1,\ldots,t_k)$ by $F^*(u,t_1,\ldots,t_k)$, and every constant symbol c by $c^*(u)$.

Case 2 Suppose $\varphi(x_{i_1},\ldots,x_{i_n})$ is $=(t_1(x_{i_1},\ldots,x_{i_n}),\ldots,t_m(x_{i_1},\ldots,x_{i_n}))$, where $i_1 < \ldots < i_n$. We define $\tau_{1,\varphi}(S)$ as follows.

Subcase 2.1 $m = 0$. We let $\tau_{1,\varphi}(U,\ W,\ S) = \forall u(u = u)$ and $\tau_{0,\varphi}(U,\ W,\ S) = \forall u \neg U(u)$.

Subcase 2.2 $m = 1$. Now $\varphi(x_{i_1},\ldots,x_{i_n})$ is $=(t_1(x_{i_1},\ldots,x_{i_n}))$. We let $\tau_{1,\varphi}$ $(U,\ W,\ S)$ be the formula

$$\forall u \forall u' \forall x_{i_1} \ldots \forall x_{i_n} \forall x_{i_{n+1}} \ldots \forall x_{i_{n+n}}((S(u,x_{i_1}\ldots x_{i_n}) \land S(u',x_{i_{n+1}}\ldots x_{i_{n+n}}))$$

$$\to t_1(x_{i_1},\ldots,x_{i_n}) = t_1(x_{i_{n+1}},\ldots,x_{i_{n+n}}))$$

and we further let $\tau_{0,\varphi}(U,\ W,\ S)$ be the formula[21] $\forall u \neg U(u)$.

---

[21] The reason for making the negation of the dependence atom false except in the empty multiverse is that the dependence atom does not have any natural negation.

Subcase 2.3 If $m > 1$ we let $\tau_{1,\varphi}(U, W, S)$ be the formula

$$\forall u \forall u' \forall x_{i_1} \ldots \forall x_{i_n} \forall x_{i_{n+1}} \ldots \forall x_{i_{n+n}}((S(u, x_{i_1} \ldots x_{i_n}) \wedge S(u', x_{i_{n+1}} \ldots x_{i_{n+n}}) \wedge$$
$$t_1(x_{i_1}, \ldots, x_{i_n}) = t_1(x_{i_{n+1}}, \ldots, x_{i_{n+n}}) \wedge$$
$$\ldots$$
$$t_{m-1}(x_{i_1}, \ldots, x_{i_n}) = t_{m-1}(x_{i_{n+1}}, \ldots, x_{i_{n+n}}))$$
$$\rightarrow t_m(x_{i_1}, \ldots, x_{i_n}) = t_m(x_{i_{n+1}}, \ldots, x_{i_{n+n}}))$$

and we further let $\tau_{0,\varphi}(U, W, S)$ be the formula $\forall u \neg U(u)$.

Case 3 Suppose $\varphi(x_{i_1}, \ldots, x_{i_n})$ is

$$= (x_{i_1} \ldots x_{i_n} : \psi(x_{i_1}, \ldots, x_{i_n})),$$

where $\psi(x_{i_1}, \ldots, x_{i_n})$ is first order. We let $\tau_{1,\varphi}(U, W, S)$ be the formula

$$\forall u, u' \forall x_{i_1} \ldots \forall x_{i_n}\{(U(u) \wedge U(u')) \rightarrow$$
$$[(W(u, x_{i_1}) \wedge \ldots \wedge W(u, x_{i_n}) \wedge (\psi^*(u, x_{i_1}, \ldots, x_{i_n}))) \leftrightarrow$$
$$(W(u', x_{i_1})) \wedge \ldots \wedge W(u', x_{i_n}) \wedge (\psi^*(u' x_{i_1}, \ldots, x_{i_n})))]\}$$

where $\varphi^*(u, x_{i_1}, \ldots, x_{i_n})$ is obtained from $\varphi(x_{i_1}, \ldots, x_{i_n})$ by replacing every predicate symbol $R(t_1, \ldots, t_k)$ by $R^*(u, t_1, \ldots, t_k)$, every function symbol $F(t_1, \ldots, t_k)$ by $F^*(u, t_1, \ldots, t_k)$, every constant symbol c by $c^*(u)$, and restricting every quantifier to $W(u, \cdot)$. We further let $\tau_{0,\varphi}(U, W, S)$ be the formula $\forall u \neg U(u)$.

Case 4 Suppose φ is NE. We let $\tau_{1,\varphi}(U, W, S)$ be the formula $\exists u U(u)$ and we further let $\tau_{0,\varphi}(U, W, S)$ be the formula $\forall u \neg U(u)$.

Case 5 Suppose $\varphi(x_{i_1}, \ldots, x_{i_n})$ is the disjunction

$$\psi(x_{j_1}, \ldots, x_{j_p}) \vee \theta(x_{k_1}, \ldots, x_{k_q}),$$

where $\{i_1, \ldots, i_n\} = \{j_1, \ldots, j_p\} \cup \{k_1, \ldots, k_q\}$. We let the sentence $\tau_{1,\varphi}(U, W, S)$ be

$$\exists U_1 \exists U_2 \exists W_1 \exists W_2 \exists S_1 \exists S_2[\Theta_p(U_1, W_1, S_1) \wedge \Theta_q(U_2, W_2, S_2) \wedge$$
$$\tau_{1,\psi}(U_1, W_1, S_1) \wedge \tau_{1,\theta}(U_2, W_2, S_2) \wedge$$
$$\forall u(U(u) \leftrightarrow (U_1(u) \vee U_2(u))) \wedge$$
$$\forall u \forall x \wedge_{i=1}^{2}(W_i(u, x) \leftrightarrow (W(u, x) \wedge U_i(u))) \wedge$$
$$\forall u \forall x_{i_1} \ldots \forall x_{i_n}(S(u, x_{i_1}, \ldots, x_{i_n}) \rightarrow$$
$$(S_1(u, x_{j_1}, \ldots, x_{j_p}) \vee S_2(u, x_{k_1}, \ldots, x_{k_q})))$$

and we let the sentence $\tau_{0,\varphi}(U, W, S)$ be

$$\exists S_1 \exists S_2 [\Theta_p(U, W, S_1) \wedge \Theta_q(U, W, S_2) \wedge$$
$$\tau_{0,\psi}(U, W, S_1) \wedge \tau_{0,\theta}(U, W, S_2)$$
$$\forall u \forall x_{i_1} \dots \forall x_{i_n}(S(u, x_{i1}, \dots, x_{i_n}) \rightarrow$$
$$(S_1(u, x_{j_1}, \dots, x_{j_p}) \wedge S_2(u, x_{k_1}, \dots, x_{k_q})))].$$

Case 6 Conjunction is handled as disjunction.

Case 7 Suppose $\varphi(x_{i_1}, \dots, x_{i_n})$ is the Boolean disjunction

$$\psi(x_{j_1}, \dots, x_{j_p}) \vee_B \theta(x_{k_1}, \dots, x_{k_q}),$$

where $\{i_1, \dots, i_n\} = \{j_1, \dots, j_p\} \cup \{k_1, \dots, k_q\}$. We let the sentence $\tau_{1,\varphi}(U, W, S)$ be

$$\exists S_1 \exists S_2 [\Theta_p(U, W, S_1) \wedge \Theta_q(U, W, S_2) \wedge$$
$$(\tau_{0,\psi}(U, W, S_1) \vee \tau_{0,\theta}(U, W, S_2))$$
$$\forall u \forall x_{i_1} \dots \forall x_{i_n}(S(u, x_{i_1}, \dots, x_{i_n}) \rightarrow$$
$$(S_1(u, x_{j_1}, \dots, x_{j_p}) \wedge S_2(u, x_{k_1}, \dots, x_{k_q})))].$$

We further let $\tau_{0,\varphi}(U, W, S)$ be the formula $\forall u \neg U(u)$.

Case 8 φ is $\neg \psi$. $\tau_{d,\varphi}(U, W, S)$ is the formula $\tau_{1-d,\psi}(U, W, S)$.

Case 9 Suppose $\varphi(x_{i_1}, \dots, x_{i_n})$ is the formula $\exists x_{i_{n+1}} \psi(x_{i_1}, \dots, x_{i_{n+1}})$.
Then $\tau_{1,\varphi}(U, W, S_1)$ is the formula

$$\exists S_1(\tau_{1,\psi}(U, W, S_1) \wedge \Theta_{n+1}(U, W, S_1) \wedge \forall u \forall x_{i_1} \dots \forall x_{i_n}(S(u, x_{i_1}, \dots, x_{i_n})$$
$$\rightarrow \exists x_{i_{n+1}}(S_1(u, x_{i_1}, \dots, x_{i_{n+1}}))))$$

and $\tau_{0,\varphi}(U, W, S)$ is the formula

$$\exists U_1 \exists W_1 \exists S_1 \exists F[\tau_{0,\psi}(U_1, W_1, S_1) \wedge \Theta_{n+1}(U_1, W_1, S_1) \wedge$$
$$\forall u \forall x((U(u) \wedge W(u, x)) \leftrightarrow U_1(F(u, x))) \wedge$$
$$\forall u \forall u' \forall x \forall x'(F(u, x) = F(u', x') \rightarrow (u = u' \wedge x = x')) \wedge$$
$$\forall u \forall x((U(u) \wedge W(u, x)) \leftrightarrow U_1(F(u, x))) \wedge$$
$$\forall x(U_1(x) \rightarrow \exists u \exists y(x = F(u, y)) \wedge$$
$$\forall z \forall u \forall x(S_1(F(u, z), x_1, \dots, x_{n+1}) \leftrightarrow (S(u, x_1, \dots, x_n) \wedge x_{n+1} = z))].$$

Case 10 The universal quantifier is handled as the existential one.

It is now straightforward to prove the equivalence of the following two statements for first-order ψ.

- $T \nvDash \psi$
- The first-order theory $\{\tau_{1,\varphi}(U, W, S): \varphi \in T\} \cup \{\Theta_0(U, W, S)\} \cup \{\exists u U(u)\} \cup \{\forall u(U(u) \rightarrow \neg \psi*(u))\}$, where $\psi*(u)$ is obtained from ψ by replacing every predicate symbol $R(t_1, \ldots, t_k)$ by $R^*(u, t_1, \ldots, t_k)$, every function symbol $F(t_1, \ldots, t_k)$ by $F^*(u, t_1, \ldots, t_k)$, every constant symbol c by $c^*(u)$, and restricting every quantifier to $W(u, \cdot)$, has a model.

So the claim follows from the completeness theorem of first-order logic. $\square$

When the above theorem is applied to multiverse set theory we get an axiomatization of first-order consequences of our desired theory, for example any of the following.

1. ZFC: pure ZFC.
2. ZFC + $=(\varphi)$ + $\neq(\psi)$: ZFC plus "φ has a truth value but ψ is absolutely undecidable."
3. ZFC + $\varphi \perp \psi$: ZFC plus "φ is independent of ψ."
4. ZFC + $\{=(\varphi^N): \varphi$ number theoretic$\}$: ZFC plus "no independence in number theory."

In many individual cases one can show that the first-order consequences are the same as first-order consequences of ZFC. In the last case we do not escape the force of Gödel's incompleteness theorem, although it may seem so. If θ is the relevant Gödel sentence, then ZFC $+\{=(\varphi^N): \varphi$ number theoretic$\}$ has two multiverse models, one with θ^N and another with $\neg\theta^N$. There is no contradiction with the fact that both satisfy $=(\theta^N)$.

We can further use dependence sentences such as $=(x : On(x))$ to say in set theory that all the universes have the same ordinals. Furthermore, there is a dependence sentence Θ_{wf} which essentially says that the ordinals are non-well-founded. Thus for first-order φ:

$$\text{ZFC}+ =(x : On(x)) \vDash \Theta_{wf} \vee \varphi$$

if and only if φ is true in all well-founded models of *ZFC*. This shows[22] that we cannot hope to axiomatize entire MD.

Finally we may adopt the ultimate uniformization axiom

$$=(x, y : x \in y)$$

which in multiverse set theory says that, after all, there is just one universe.

---

[22] And there are stronger results, reducing the truth of a Π_2-sentence (in V) to consequence in MD.

5.3   The generic multiverse

We shall now show that we can capture the generic multiverse of Woodin and Steel with multiverse dependence logic. Notice that truth in their generic multiverse can even be captured by first-order logic in the sense that truth of a given sentence φ in the generic multiverse can be expressed as the truth of another sentence $\varphi *$ in V. So multiverse dependence logic is not needed in this case. However, the method by which we characterize the generic multiverse in MD is so general that it applies to any similar situation.

We first recall an interesting result of Laver (2007, Theorem 3). There is a formula $\varphi(x, y)$ of set theory with the following property. Suppose P is a forcing notion, $\delta = |P|$, and G is P-generic over V. Then in $V[G]$ the formula $\varphi(x, y)$ defines the ground model V with the set $V_{\delta+1}$ as a parameter, that is, in $V[G]$ the following holds:

$$V = \{a : \varphi(a, V_{\delta+1})\}.$$

The formula $\varphi(x, y)$ is by no means trivial, but it can be explicitly written down. Note that although in V the universe is trivially definable by the formula $x = x$, after the forcing by P the old universe may a priori be completely hidden.

We now define a logical operation GMV (really a new atomic formula) so that in the context of multiverse set theory we have the following.

$\mathbb{M} \models$ GMV if and only if the following are true.

- For any universe M in $\mathbb{M}$ and any poset P in M there is a generic extension of M by P in $\mathbb{M}$.
- For any universe M in $\mathbb{M}$ and any poset P in M, if M is a generic extension of N by P, then $N \in \mathbb{M}$.
- For any universes $M, M' \in \mathbb{M}$ there are generic extensions $M[G]$ and $M'[G']$ of M and M' in $\mathbb{M}$ such that $M[G] = M'[G']$.

The logical operation GMV can be added to MD without losing Theorem 5.6; but in the case of ZFC we do not get any new first-order consequences. The conditions in the definition of GMV are the conditions that characterize Steel's generic multiverse (Chapter 8). So adding GMV to the ZFC axioms means in multiverse set theory the same as restricting the multiverse to the generic multiverse generated by V.

6   Conclusion

The working mathematician need not worry whether he or she is working in a one universe setup or a multiverse setup, because the two, as I have explained to them, are in harmony with each other.

But if the mathematician wants to incorporate in his or her investigation the firm conviction that a certain proposition has a determined truth-value, although this conviction does not lead to a conclusion as to whether the true-value is true or false, he or she can use the operation $=(\varphi)$ to add a new axiom to this effect. Respectively, if a mathematician has a firm conviction that a certain proposition lacks a truth-value, i.e., is absolutely undecidable, he or she can use the operation $\neq(\varphi)$ to add an axiom to this effect.

The sentences $=(\varphi)$ and $\neq(\varphi)$ are examples of sentences in a new *multiverse dependence logic* which provides a whole arsenal of methods to inject order into the multiverse.

PART V

The legacy

CHAPTER 10

Undecidable problems: a sampler

Bjorn Poonen

1 Introduction

The goal of this chapter is to demonstrate that undecidable decision problems arise naturally in many branches of mathematics. The criterion for selection of a problem in this survey is simply that the author finds it entertaining! We do not pretend that our list of undecidable problems is complete in any sense. And some of the problems we consider turn out to be decidable or to have unknown decidability status. For another survey of undecidable problems, see Davis (1977).

2 Two notions of undecidability

There are two common senses in which one speaks of undecidability.

1. **Independence from axioms:** A single statement is called undecidable if neither it nor its negation can be deduced using the rules of logic from the set of axioms being used. (Example: The continuum hypothesis, that there is no cardinal number strictly between $\aleph_0$ and $2^{\aleph_0}$, is undecidable in the ZFC axiom system, assuming that ZFC itself is consistent (Gödel 1940; Cohen 1963, 1964).) The first examples of statements independent of a "natural" axiom system were constructed by Gödel (1931b).

2. **Decision problem:** A family of problems with YES/NO answers is called undecidable if there is no algorithm that terminates with the correct answer for every problem in the family. (Example: Hilbert's tenth problem, to decide whether a multivariable polynomial equation with integer coefficients has a solution in integers, is undecidable (Matiyasevich 1970).)

I thank Henry Cohn, Martin Davis, Jeffrey Lagarias, and Richard Stanley for discussions. The writing of this chapter was supported by the Guggenheim Foundation and National Science Foundation grants DMS-0841321 and DMS-1069236.

Remark 2.1 In modern literature, the word "undecidability" is used more commonly in sense 2, given that "independence" adequately describes sense 1.

To make sense 2 precise, one needs a formal notion of algorithm. Such notions were introduced by Church (1936a) and Turing (1936) independently in the 1930s. From now on, we interpret algorithm to mean Turing machine, which, loosely speaking, means that it is a computer program that takes as input a finite string of 0s and 1s. The role of the finite string is to specify which problem in the family is to be solved.

Remark 2.2 Often in describing a family of problems, it is more convenient to use higher-level mathematical objects such as polynomials or finite simplicial complexes as input. This is acceptable if these objects can be encoded as finite binary strings. It is not necessary to specify the encoding as long as it is clear that a Turing machine could convert between reasonable encodings imagined by two different readers.

Remark 2.3 One cannot speak of a *single* YES/NO question being undecidable in sense 2, because there exists an algorithm that outputs the correct answer for it, even if one might not know *which* algorithm it is!

There is a connection between the two notions of undecidability. Fix a decision problem and an axiom system $\mathcal{A}$ such that

(a) there is a computer program that generates exactly the axioms of $\mathcal{A}$; and

(b) there is a computer program that, when fed an instance i of the decision problem, outputs a statement Y_i in the language of $\mathcal{A}$ such that

  • if Y_i is provable in $\mathcal{A}$, then the answer to i is YES, and
  • if $\neg\, Y_i$ is provable in $\mathcal{A}$, then the answer to i is NO.

Under these assumptions, if the decision problem is undecidable in sense 2, then at least one of its instance statements Y_i is undecidable in sense 1, i.e., independent of $\mathcal{A}$. The proof of this is easy: if every Y_i could be proved or disproved in $\mathcal{A}$, then the decision problem could be solved by a computer program that generates all theorems deducible from $\mathcal{A}$ until it finds either Y_i or $\neg\, Y_i$. In fact, under the same assumptions, there must be *infinitely many* Y_i that are independent of $\mathcal{A}$, since if there were only finitely many, there would exist a decision algorithm that handled them as special cases.

Remark 2.4 In all the undecidable decision problems we present, the source of the undecidability can be traced back to a single undecidable decision problem, namely the halting problem, or equivalently the membership problem for listable sets (see Sections 3.1 and 3.2). For any of these problems, in principle we can compute a *specific i* for which Y_i is independent of $\mathcal{A}$ (cf. Post 1944, the last paragraph of p. 294). The value of i depends on $\mathcal{A}$; more precisely, i can be computed in terms of the programs in (a) and (b) above.

Example 2.5 Assume that ZFC is consistent and, moreover, that theorems in ZFC about integers are true. Then, because the undecidability of Hilbert's tenth problem in sense 2 is proved via the halting problem (see Section 8.1), there is a specific polynomial $f \in \mathbb{Z}[x_1, \ldots, x_n]$ one could write down in principle such that neither

$$(\exists x_1, \ldots, x_n \in \mathbb{Z}) \, f(x_1, \ldots, x_n) = 0 \tag{1}$$

nor its negation can be proved in ZFC. Moreover, (1) must be false, because if it were true, it could be proved in ZFC by exhibiting a single $(x_1, \ldots, x_n) \in \mathbb{Z}^n$ satisfying $f(x_1, \ldots, x_n) = 0$. (It might seem as if this is a ZFC proof of the negation of (1), but in fact it is only a ZFC proof of the *implication*

> If ZFC is consistent and proves only true theorems about integers, then the negation of (1) holds.

This observation is related to Gödel's second incompleteness theorem, which implies that ZFC cannot prove the hypothesis of the implication unless ZFC is inconsistent!)

3 Logic

Gödel's incompleteness theorems (Gödel 1931b) provided undecidable statements in sense 1 for a wide variety of axiom systems. Inspired by this, Church and Turing began to prove that certain decision problems were undecidable in sense 2, as soon as they developed their notions of algorithm.

3.1 The halting problem

The halting problem asks whether it is possible to write a debugger that takes as input a computer program and decides whether it eventually halts instead of entering an infinite loop. For convenience, let us assume that each program accepts a natural number as input:

Halting problem
> **input:** a program p and a natural number x
> **question:** Does p eventually halt when run on input x?

Theorem 3.1 (Turing 1936) *The halting problem is undecidable.*

Sketch of proof. We will use an encoding of programs as natural numbers, and identify programs with numbers. Suppose that there were an algorithm for deciding when program p halts on input x. Using this, we could write a new program H such that

H halts on input x $\Leftrightarrow$ program x does not halt on input x.

Taking $x = H$, we find a contradiction: H halts on input H if and only if H does not halt on input H.

To turn the sketch above into a complete proof would require some programming, to show that there is a "universal" computer program that can simulate any other program given its number; this could then be used to construct H.

3.2 Listable sets

Let $\mathbb{N}$ be the set of natural numbers. Let A be a subset of $\mathbb{N}$. Call A computable[1] if there is an algorithm that takes as input an element $n \in \mathbb{N}$ and decides whether or not $n \in A$. On the other hand, call A listable or computably enumerable (c.e.) if there is a computer program that when left running forever eventually prints out exactly the elements of A. Computable sets are listable.

For each listable set A, we then have the following decision problem:

Membership in a listable set A
> **input:** $n \in \mathbb{N}$
> **question:** Is $n \in A$?

Theorem 3.2 (Church 1936a; Kleene 1936; Rosser 1936) *There exists a listable set A for which the membership problem is undecidable.*

---

[1] In most twentieth century literature on the subject one finds the terms recursive and recursively enumerable (r.e.). But Soare (1996) has argued in favor of the use of the terms "computable" and "c.e." instead, and many researchers in the field have followed his recommendation.

Proof. Let A be the set of numbers of programs that halt. Then A is listable (write a program that during iteration N runs each of the first N programs for N steps, and prints the numbers of those that have already halted). But the undecidability of the halting problem implies that A is not computable; in other words, the membership problem for A is undecidable.

It would be just as easy to argue in reverse, to use the existence of a noncomputable listable set to prove the undecidability of the halting problem.

3.3   The Entscheidungsproblem

Fix a finite set of axioms. Then there are some (first-order) statements that are universally valid, meaning that they are true for every mathematical structure satisfying the axioms. By the completeness theorem of first-order logic (Gödel 1930c), the universally valid statements are exactly the ones that are provable in the sense that they can be deduced from the axioms using the rules of logic.

Can one decide in a finite amount of time whether or not any given statement is universally valid? This is the Entscheidungsproblem, proposed by Hilbert (Hilbert and Ackermann 1928, Chapter 3, §11). (*Entscheidung* is the German word for "decision.") One could try searching for a proof by day and searching for a proof of the negation by night, but such an algorithm might fail to terminate for some input statements since it could be that neither proof exists.

More formally, but still without providing full definitions, given a first-order logic $\mathcal{F}$, possibly including a finite number of special axioms beyond the basic axioms of first-order logic, one has the following decision problem:

Entscheidungsproblem for $\mathcal{F}$

>   **input:** a first-order sentence s in the language of $\mathcal{F}$
>   **question:** Is s true in every model of the axioms of $\mathcal{F}$?

It was known to Hilbert that there is a single first-order logic $\mathcal{F}_0$ without special axioms such that if the Entscheidungsproblem for $\mathcal{F}_0$ is decidable, so is the Entscheidungsproblem for any other first-order logic. But Church (1936a, 1936b) and Turing (1936, §11) independently proved that the Entscheidungsproblem for $\mathcal{F}_0$ was undecidable. For more information, see Davis (1958, Chapter 8, §4).

4 Combinatorics

4.1 The Post correspondence problem

Imagine a rectangular block with a finite string of *a*s and *b*s written along the top and another such string written along the bottom, both upright. When finitely many such blocks are laid side to side, the strings along the top may be concatenated, and the strings along the bottom may be concatenated. Post (1946) proved that the following simple-sounding problem is undecidable.

> Post correspondence problem
> > **input:** a finite collection of blocks, labeled as above
> > **question:** Given an unlimited supply of copies of these particular blocks, can one form a nonempty finite sequence of them for which the concatenation of the top strings equals the concatenation of the bottom strings?

The reason that this is undecidable is that one can embed the halting problem in it. Namely, with some work it is possible, given a computer program p, to construct an instance of the Post correspondence problem that has a positive answer if and only if p halts.

Because of its simplicity, the Post correspondence problem is often used to prove the undecidability of other problems, for instance, in the formal theory of languages (see Davis 1977).

4.2 Tiling the plane

Wang tiles, introduced by H. Wang (1961, §4.1), are unit squares in the plane, with sides parallel to the axes, such that each side of each square has been assigned a color. They may be translated, but not rotated or reflected. A tiling of the plane into such squares is valid if whenever two squares share an edge, the colors match, as in the game of dominoes. Wang proposed the following problem.

> Tiling problem
> > **input:** a finite collection of Wang tiles
> > **question:** Is there a valid tiling of the entire plane using only translated copies of the given tiles?

Wang (1961, 4.1.2) also conjectured that if a tiling exists for a given finite collection, then there exists a *periodic* tiling, i.e., one that is invariant under translations by the vectors in a finite-index subgroup of $\mathbb{Z}^2$, or equivalently by the vectors in $(n\mathbb{Z})^2$ for some fixed $n \geq 1$. He observed that this conjecture would imply that the tiling problem was decidable: on the nth day one could search for tilings that are invariant under translations in $(n\mathbb{Z})^2$, and on the nth night one could search for an $n \times n$ square that cannot be tiled (a compactness argument shows that if the entire plane cannot be tiled, then there exists n such that the $n \times n$ square cannot be tiled).

But Berger (1966) then proved that the tiling problem was undecidable, by embedding the halting problem as a subproblem of the tiling problem. Combining this with Wang's observation shows that there exist finite collections that can tile the plane, but only *aperiodically*. Simplifications by Robinson (1971), Kari (1996), and Culik (1996) led to an example with only 13 tiles.

Remark 4.1 Robinson (1978) and Margenstern (2008) proved similar undecidability results for tilings of the *hyperbolic* plane.

Other tiling problems involve polyominoes. A polyomino is a connected planar region obtained by connecting finitely many unit squares along shared edges. It is unknown whether the following is undecidable (see Rhoads 2005, p. 330, for instance).

Polyomino tiling
> **input:** a polyomino P
> **question:** Can one tile the entire plane using translated and rotated copies of P?

4.3 Graph theory

Fix finite graphs G and H. Let $V(G)$ be the vertex set of G; define $V(H)$ similarly. A homomorphism from H to G is a (not necessarily injective) map $V(H) \to V(G)$ such that every edge of H maps to an edge of G. The homomorphism density $t(H,G)$ is the probability that a uniformly chosen random map $V(H) \to V(G)$ is a homomorphism. If $H_1 \dot\cup H_2$ denotes the disjoint union of graphs H_1 and H_2, then $t(H_1 \dot\cup H_2, G) = t(H_1, G)t(H_2, G)$ for any G.

There are certain known inequalities relating these densities. For instance, for the complete graph K_n on n vertices, elementary counting arguments similar to those in Goodman (1959) show that

$$t(K_3, G) \geq 2t(K_2, G)^2 - t(K_2, G),$$

or equivalently

$$t(K_3, G) - 2t(K_2 \dot\cup K_2, G) + t(K_2, G) \geq 0,$$

for every finite graph G. This suggests the following problem.

Linear inequalities between graph homomorphism densities

> **input:** $k \in \mathbb{Z}_{\geq 0}$, finite graphs $H_1, \ldots, H_k$, and integers $a_1, \ldots, a_k$
> **question:** Does $a_1 t(H_1, G) + \cdots + a_k t(H_k, G) \geq 0$ hold for all finite graphs G?

Hatami and Norine (2011, Theorem 2.12) proved this problem undecidable by relating it to Hilbert's tenth problem.

5 Matrix semigroups

5.1 Matrix mortality

Given a finite list of square integer matrices, there are many ways to form products, especially if the factors may be repeated. Can one decide whether some product yields the zero matrix **0**? More formally, we have the following.

Matrix mortality problem

> **input:** $n \in \mathbb{Z}_{\geq 0}$ and a finite set S of $n \times n$ integer matrices
> **question:** Does the multiplicative semigroup generated by S contain **0**?

Paterson (1970) proved that this problem is undecidable, even for sets of 3×3 matrices, via reduction to the Post correspondence problem. Subsequent work showed that it is undecidable also for sets consisting of seven 3×3 matrices (Halava *et al.* 2007, Corollary 1) and for sets consisting of two 21×21 matrices (Halava *et al.* 2007, Theorem 11). Whether there exists an algorithm for sets of 2×2 matrices remains an open problem. For a more detailed introduction to the matrix mortality problem, see Halava and Harju (2001).

5.2 Freeness

One can ask, given n and S, whether distinct finite sequences of matrices in S yield distinct products, i.e., whether the semigroup generated by S is free. This turns out to be undecidable too, and already for sets of 3×3 matrices (Klarner *et al.* 1991). In fact, sets of fourteen 3×3 matrices suffice for undecidability (Halava *et al.* 2007, Theorem 13).

5.3 Finiteness

Can one decide whether the semigroup generated by S is finite? This time the answer turns out to be yes, as was proved independently by Jacob (1977–1978, 1978) and by Mandel and Simon (1977–1978).

Let us outline a proof. The main step consists of showing that there is a computable bound $f(n, s)$ for the size of any finite semigroup of $M_n(\mathbb{Z})$ generated by s matrices. Now for any $r \geq 1$, let P_r be the set of products of length at most r of matrices in S. Start computing P_1, P_2, and so on. If $\#P_1 < \cdots < \#P_N$ for $N = f(n, s) + 1$, then $\#P_N > f(n, s)$, so the semigroup is infinite. Otherwise $P_r = P_{r+1}$ for some $r < N$, in which case the semigroup equals P_r and hence is finite.

The algorithm can be extended to decide finiteness of a finitely generated semigroup of $M_n(k)$ for any finitely generated field k presented as an explicit finite extension of $\mathbb{F}_p(t_1, \ldots, t_d)$ or $\mathbb{Q}(t_1, \ldots, t_d)$.

5.4 Powers of a single matrix

There are even some nontrivial questions about semigroups generated by one matrix! Given $A \in M_k(\mathbb{Z})$, can one decide whether there exists $n \in \mathbb{Z}_{>0}$ such that the upper right corner of A^n is 0? This problem, whose undecidability status is unknown, is equivalent to the following.

Zero in a linear recursive sequence

> **input:** a linear recursive sequence of integers $(x_n)_{n \geq 0}$, specified by giving $x_0, \ldots, x_{k-1} \in \mathbb{Z}$ and $a_0, \ldots, a_{k-1} \in \mathbb{Z}$ such that $x_{n+k} = a_{k-1}x_{n+k-1} + \cdots + a_0x_n$ for all $n \geq 0$
> **question:** Does there exist n such that $x_n = 0$?

This is known also as Skolem's problem, since Skolem (1934) proved that $\{n: x_n = 0\}$ is a union of a finite set and finitely many arithmetic progressions. See Halava *et al.* (2005).

6 Group theory

Motivated by topology, Dehn (1911) asked three questions about groups.

1. Is there an algorithm to recognize the identity of a group?
2. Is there an algorithm to decide whether two given elements of a group are conjugate?

3. Is there an algorithm to decide whether two given groups are isomorphic?

Dehn formulated the questions precisely, except for the precise notion of algorithm.

6.1 Finitely presented groups

To make sense of such questions, one must specify how a group is presented and how an element is presented. A natural choice is to describe a group by means of a finite presentation such as

$$S_3 = \langle r, t : r^3 = 1, t^2 = 1, trt^{-1} = r^{-1} \rangle.$$

This example describes the group of symmetries of an equilateral triangle as being generated by a 120° rotation r and a reflection t, and lists relations satisfied by r and t such that all other relations are consequences of these. More formally, if $n \in \mathbb{Z}_{\geq 0}$, and F_n is the free group on n generators, and R is a finite subset of F_n, and H is the smallest normal subgroup of F_n containing R, then we may form the quotient group F_n/H. Any group arising in this way is called a finitely presented (f.p.) group. An element of an f.p. group can be specified by giving a word in the generators, i.e., a finite sequence of the generators and their inverses, such as $rtr^{-1}r^{-1}ttt^{-1}$.

6.2 The word problem

For each fixed f.p. group G (or more precisely, for each such group equipped with a particular presentation), we have the following.

Word problem for an f.p. group G
    input: word w in the generators of G
    question: Does w represent 1 in G?

The decidability of the word problem depends only on the isomorphism type of the group, and not on the presentation. There are many classes of groups for which the word problem is decidable: finite groups, f.p. abelian groups, and free groups on finitely many generators, for instance. (For free groups, one algorithm is to cancel pairs of adjacent inverse symbols repeatedly for as long as possible; the resulting reduced word represents 1 if and only if it is empty.)

But in the 1950s, P.S. Novikov (1955) and Boone (1959) independently proved that there is an f.p. group for which the word problem is undecidable. The analogue for f.p. semigroups had been proved earlier, by Post (1947) and Markov (1947, 1951); one proof of this goes through the undecidability of another word problem, namely that for semi-Thue systems, which can also be used to prove undecidability of the Post correspondence problem. Ultimately, the proofs of all these results are via reduction to the halting problem. Novikov and Boone essentially showed that for a certain f.p. group G, one could associate to any computer program p a word w in the generators of G such that w represents $1 \in G$ if and only if p halts.

The undecidability of the word problem admits another proof, using the Higman embedding theorem, which we state below after introducing a definition. A finitely generated group is called recursively presented if it has the form F_n/H, where H is the smallest normal subgroup of F_n containing a given subset R, which is no longer required to be finite, but is instead required to be listable. Amazingly, it is possible to characterize such groups without mentioning computability.

Higman embedding theorem (Higman 1961) *A finitely generated group is recursively presented if and only if it can be embedded in a finitely presented group.*

. The Higman embedding theorem implies the existence of an f.p. group P with undecidable word problem, as we now explain. First, it is rather easy to construct a recursively presented group for which the word problem is undecidable: for instance, if S is any noncomputable listable set of positive integers, then one can show that in the recursively presented group

$$G_S := \langle a, b, c, d \,|\, a^n b a^{-n} = c^n d c^{-n} \text{ for all } n \in S \rangle,$$

$a^n b a^{-n} c^n d^{-1} c^{-n}$ represents 1 if and only if $n \in S$, so G_S has an undecidable word problem. By the Higman embedding theorem, G_S embeds in some finitely presented group P, which therefore has an undecidable word problem too.

6.3 *The conjugacy problem*

For each fixed f.p. group G, we have another problem:

Conjugacy problem for an f.p. group G
    input: words w_1, w_2 in the generators of G
    question: Do w_1 and w_2 represent conjugate elements of G?

The word problem can be viewed as the subproblem of the conjugacy problem consisting of the instances for which w_2 is the empty word, which represents 1. Thus the conjugacy problem for G is at least as hard as the word problem for G, which means that it is easier (or at least no harder) to find a G for which the conjugacy problem is undecidable. In fact, Novikov (1954) published a proof of the existence of an f.p. group for which the conjugacy problem is undecidable before publishing the result on the word problem, and this earlier proof is much simpler.

The inequality above between the difficulties of the two problems is the only one, in a sense that can be made precise using basic notions of computability theory, namely the notions of c.e. degrees of unsolvability and Turing reducibility $\leq_T$:

Theorem 6.1 (Collins 1972) *Given c.e. degrees W and C such that $W \leq_T C$, there exists an f.p. group G for which the word problem has degree W and the conjugacy problem has degree C.*

This means that given c.e. subsets W and C of $\mathbb{N}$ such that the membership problem for W is decidable given an oracle for the membership problem for C, there exists an f.p. group G such that the word problem for G can be solved using an oracle for membership in W and vice versa, and the conjugacy problem for G can be solved using an oracle for membership in C and vice versa.   •

6.4   Properties of groups

Instead of fixing an f.p. group G, one can ask about algorithms that accept a finite presentation as input and try to decide whether the group it defines has a given property. For a wide variety of natural properties, the decision problem turns out to be undecidable. To make this precise, define a Markov property to be a property P of f.p. groups, depending only on the isomorphism type of the group, not on the presentation, such that

1.   there exists an f.p. group G_1 with P, and
2.   there exists an f.p. group G_2 that cannot be embedded in any f.p. group with P.

Examples are the properties of being trivial, finite, abelian, nilpotent, solvable, free, or torsion-free, because all these properties are inherited by subgroups. The property of having a decidable word problem is yet another Markov property, for the same reason!

Using the undecidability of the word problem, Adian (1957a, 1957b) and Rabin (1958) proved the following.

Theorem 6.2 *For any Markov property P, it is impossible to decide whether an f.p. group G has P.*

Corollary 6.3 *It is impossible to decide whether a finite presentation describes the trivial group.*

Deciding triviality is a subproblem of the general problem of deciding whether two finite presentations define isomorphic groups, so the isomorphism problem is undecidable too.

For a more extended survey of undecidability in group theory, see Miller (1992).

7 Topology

7.1 *The homeomorphism problem*

Given two manifolds, can one decide whether they are homeomorphic? As usual, to make sense of such a question, we need to specify how a manifold is described. Since every compact smooth manifold can be triangulated, a natural choice is to use finite simplicial complexes to represent manifolds.

> Homeomorphism problem
>
> **input:** finite simplicial complexes M and N representing smooth manifolds
>
> **question:** Are M and N homeomorphic?

(One could alternatively replace homeomorphic by PL-homeomorphic, where PL stands for piecewise-linear.)

Given a finite simplicial complex M representing a compact manifold, one obtains a subproblem of the homeomorphism problem by fixing the first input to be M:

> Recognizing M
>
> **input:** a finite simplicial complex N representing a smooth manifold
>
> **question:** Is N homeomorphic to M?

One can also restrict these problems according to dimension. For $d \le 3$, the homeomorphism problem for d-folds turns out to be decidable, because of classification theorems; for $d = 3$, this uses the work of Perelman on Thurston's geometrization conjecture. But for each $d \ge 4$, the homeomorphism problem for d-folds is undecidable, as was proved by Markov (1958). Moreover, S. P. Novikov (the son of P. S. Novikov!) proved that recognizing whether a finite simplicial complex is homeomorphic to the d-sphere S^d is an undecidable problem for each $d \geq 5$ (a proof appears in the appendix to Volodin *et al.* (1974)). From this, one can prove that for any fixed compact d-fold M with $d \geq 5$, recognizing whether a finite simplicial complex is homeomorphic to M is undecidable.

All these results are proved by reduction to the undecidability results for f.p. groups. We now sketch the proofs of the unrecognizability results. (For a survey with more details, see Weinberger (2005, Chapter 2).) Fix $d \geq 5$. Choose an f.p. group G with undecidable word problem. From G and a word w in the generators of G, one can build an f.p. group G_w such that G_w is trivial if and only if w represents 1, and such that the first and second homology groups $H_1(G_w)$ and $H_2(G_w)$ are trivial. These conditions on H_1 and H_2 of an f.p. group are necessary and sufficient for there to exist a homology sphere (a compact d-manifold with the same homology as S^d) with that fundamental group. In fact, one can effectively construct a finite simplicial complex X_w representing such a homology sphere with fundamental group G_w. Now we have the following.

- If w represents 1, then G_w is trivial, and X_w is a simply connected homology sphere, but in dimensions $d \geq 5$ a theorem of Smale (1961) implies that any such space is homeomorphic to S^d.
- If w does not represent 1, then X_w has nontrivial fundamental group, so X_w is not homeomorphic to S^d.

Hence, if we had an algorithm to recognize whether a finite simplicial complex is homeomorphic to S^d, it could be used to solve the word problem for G, a contradiction. Thus recognizing S^d is an undecidable problem.

Next suppose that M is *any* compact d-fold for $d \geq 5$. The connected sum $M\#X_w$ is obtained by punching a small hole in each of M and X_w and connecting them with a thin cylinder. This construction can be done effectively on finite simplicial complexes. The fundamental group $\pi_1(M\#X_w)$ is the free product of the groups $\pi_1(M)$ and $\pi_1(X_w)$. A group-theoretic theorem states that a free product $G * H$ of finitely generated groups can be isomorphic to G only if H is trivial. Now we have the following.

- If w represents 1, then X_w is homeomorphic to S^d, and $M\#X_w$ is homeomorphic to M.
- If w does not represent 1, then $M\#X_w$ does not even have the same fundamental group as M.

Hence, if we had an algorithm to recognize M, it could be used to solve the word problem for G, a contradiction.

Question 7.1 Is S^4 recognizable?

Remark 7.2 Seidel used similar ideas to find undecidable problems in *symplectic* geometry (see Seidel 2008, Corollary 6.8).

7.2 Am I a manifold?

We have seen that it is impossible to recognize whether two manifolds represented by given finite simplicial complexes are homeomorphic. Even worse, one cannot even decide whether a finite simplicial complex represents a manifold! In other words, the following problem is undecidable.

  Manifold detection

input: finite simplicial complex M
question: Is M homeomorphic to a manifold?

Let us prove the undecidability by embedding the word problem in this problem. Recall that in Section 7.1, we constructed a finite simplicial complex X_w in terms of a word w in the generators of an f.p. group G with unsolvable word problem, such that

- if w represents 1, then X_w is homeomorphic to a sphere S^d, and
- if w does not represent 1, then X_w is a manifold with nontrivial fundamental group.

The suspension SX_w of X_w is the simplicial complex whose vertices are those of X_w together with two new points a and b, and set of faces is $\cup_{\Delta \in X_w} \{\Delta, \Delta \cup \{a\}, \Delta \cup \{b\}\}$. Geometrically, one may realize X_w in a hyperplane in $\mathbb{R}^n$, and a and b as points on either side of the hyperplane; then SX_w is the union of the line segments connecting a point of $\{a, b\}$ to a point of the realization of X_w. Now we have the following.

- If w represents 1, then X_w is homeomorphic to a sphere S^d, and SX_w is homeomorphic to a sphere S^{d+1}.
- If w does not represent 1, then X_w has nontrivial fundamental group, so SX_w contains loops arbitrarily close to a with nontrivial class in the fundamental group of $SX_w - \{a, b\}$, so SX_w is not locally Euclidean at a.

Thus SX_w is homeomorphic to a manifold if and only if w represents 1. Therefore no algorithm can decide whether a given finite simplicial complex represents a manifold.

7.3   Knot theory

A knot is a smooth embedding of the circle S^1 in $\mathbb{R}^3$. Two knots are equivalent if there is an ambient isotopy that transforms one into the other; loosely speaking, this means that there is a smoothly varying family of diffeomorphisms of $\mathbb{R}^3$, parametrized by an interval, starting with the identity and ending with a diffeomorphism that maps one knot onto the other.

How do we describe a knot in a way suitable for input into a computer? A knot may be represented by a finite sequence of distinct points in $\mathbb{Q}^3$: the knot is obtained by connecting the points in order by line segments, the last of which connects the last point back to the first point (we assume that each segment intersects its neighbors only at its endpoints and intersects other segments not at all, and the piecewise-linear curve should then be rounded at the vertices so as to obtain a smooth curve).

> Knot equivalence problem
>   **input:** knots K_1 and K_2, each represented by a finite sequence in $\mathbb{Q}^3$
>   **question:** Are K_1 and K_2 equivalent?

Haken (1961) constructed an algorithm to decide whether a knot was unknotted, and for the general problem he outlined an approach, the last step of which was completed by Hemion (1979). Thus the knot equivalence problem is decidable!

One can also consider knots in higher dimensions. An n-dimensional knot is a smooth embedding of S^n in $\mathbb{R}^{n+2}$ (or S^{n+2}), and one can define equivalence as before. Any embedding equivalent to the standard embedding of S^n as the unit sphere in a hyperplane in $\mathbb{R}^{n+2}$ is called unknotted. Nabutovsky and Weinberger (1996) proved that the problem of deciding whether an n-dimensional knot is unknotted is undecidable for $n \geq 3$. Since this is a subproblem of the equivalence problem for n-dimensional knots, the latter is undecidable too.

Nabutovsky and Weinberger leave open the following question.

Question 7.3 Is the equivalence problem for 2-dimensional knots decidable?

See Soare (2004) for an exposition of some other undecidable problems in topology and differential geometry.

8 Number theory

8.1 Hilbert's tenth problem

One of the 23 problems in a list that Hilbert published after a famous lecture in 1900 asked for an algorithm to decide the solvability of diophantine equations.

> Hilbert's tenth problem
> **input:** a multivariable polynomial $f \in \mathbb{Z}[x_1, \ldots, x_n]$
> **question:** Does there exist $\vec{a} \in \mathbb{Z}^n$ with $f(\vec{a}) = 0$?

This was eventually proved undecidable by Matiyasevich (1970). To explain more, we need a definition. Call a subset A of $\mathbb{Z}$ diophantine if there exists a polynomial $p(t, \vec{x}) \in \mathbb{Z}[t, x_1, \ldots, x_n]$ such that

$$A = \{a \in \mathbb{Z} : (\exists \, \vec{x} \in \mathbb{Z}^n) \, p(a, \vec{x}) = 0\}.$$

In other words, if one views $p(t, \vec{x}) = 0$ as a family of diophantine equations in the variables $x_1, \ldots, x_n$ depending on a parameter t, then A is the set of parameter values that yield a solvable diophantine equation.

It is easy to see that diophantine sets are listable. What is remarkable is that the converse holds:

Theorem 8.1 (conjectured in Davis (1953, p. 35) and proved in Matiyasevich (1970)) *A subset of $\mathbb{Z}$ is diophantine if and only if it is listable.*

Work of Davis, Putnam, and Robinson culminating in Davis *et al.* (1961) proved the analogue for exponential diophantine equations, in which polynomials are replaced by expressions built up from integers using not only addition and multiplication, but also exponentiation. Matiyasevich then showed how to express exponentiation in diophantine terms, to complete the proof of Theorem 8.1.

Theorem 8.1 immediately yields a negative answer to Hilbert's tenth problem, because there are listable subsets A of $\mathbb{Z}$ for which there is no algorithm to decide whether a given integer belongs to A (see Section 3.2). The role played by Theorem 8.1 for Hilbert's tenth problem is similar to the role played by the Higman embedding theorem (Section 6.2) for the word problem.

8.2 Hilbert's tenth problem for other rings

After the negative answer to Hilbert's tenth problem, researchers turned to variants in which the ring $\mathbb{Z}$ is replaced by some other commutative ring, such as $\mathbb{Q}$, or the ring of integers $\mathcal{O}_k$ of a fixed number field.

8.2.1 The field of rational numbers

The problem for $\mathbb{Q}$ is equivalent to the problem of deciding whether an algebraic variety over $\mathbb{Q}$ has a rational point, because any variety is a finite union of affine varieties, and any system of equations $f_1(\vec{x}) = \cdots = f_m(\vec{x}) = 0$ is solvable over $\mathbb{Q}$ if and only if the single equation $f_1(\vec{x})^2 + \cdots + f_m(\vec{x})^2 = 0$ is. It is still not known whether an algorithm exists for this problem. The notion of a subset of $\mathbb{Q}$ being diophantine over $\mathbb{Q}$ can be defined as in the previous section, except with all variables running over $\mathbb{Q}$ instead of over $\mathbb{Z}$. If the subset $\mathbb{Z}$ were diophantine over $\mathbb{Q}$, then an easy reduction to Matiyasevich's theorem would prove the undecidability of Hilbert's tenth problem for $\mathbb{Q}$. Koenigsmann (2010, Corollary 2), building on Poonen (2009), proved that the *complement* $\mathbb{Q} - \mathbb{Z}$ is diophantine over $\mathbb{Q}$; a generalization to number fields was recently proved by Park (2012).

In hopes of finding an undecidable problem, one can make the problem harder, by asking for an algorithm to decide the truth of first-order sentences, such as

$$(\exists x)(\forall y)(\exists z)(\exists w) \quad (x \cdot z + 3 = y^2) \vee \neg (z = x + w).$$

Using the theory of quadratic forms over $\mathbb{Q}$, Robinson (1949) proved that the following decision problem is undecidable.

Decision problem for the first-order theory of $\mathbb{Q}$

    input: a first-order sentence ϕ in the language of fields
    question: Is ϕ true when the variables run over elements of $\mathbb{Q}$?

8.2.2 Rings of integers

Recall that a number field is a finite extension k of $\mathbb{Q}$, and that the ring of integers $\mathcal{O}_k$ of k is the set of $\alpha \in k$ satisfying $f(\alpha) = 0$ for some monic $f(x) \in \mathbb{Z}[x]$. The problem for $\mathcal{O}_k$ is conjectured to have a negative answer for each k (Denef and Lipshitz 1978). This has been proved for some k, namely when k is totally real (Denef 1980), k is a quadratic extension of a totally real number field (Denef and Lipshitz 1978), or k has exactly one conjugate pair of nonreal embeddings (Pheidas 1998; Shlapentokh 1989). Through arguments of the author and Shlapentokh (Poonen 2002, Theorem 1; Shlapentokh 2008, Theorem 1.9(3)), certain statements about ranks of elliptic curves over number fields would imply a negative answer for every k, and such statements have been proved by Mazur and Rubin (2010, §8) assuming a conjecture of Shafarevich and Tate.

For more about Hilbert's tenth problem and its variants, see the survey articles (Davis *et al.* 1976; Mazur 1994; Poonen 2008), the books (Matiyasevich 1993; Denef *et al.* 2000; Shlapentokh 2007), the website of Vserminov, and the movie (Csicsery 2008).

9 Analysis

9.1 Inequalities

Given a real-valued function on $\mathbb{R}$ or on $\mathbb{R}^n$, can one decide whether it is nonnegative everywhere? The answer depends on the kind of functions allowed as input.

9.1.1 Real polynomials

For polynomials in any number of variables, Tarski showed that the answer is yes (to make sense of this, one should restrict the input to have coefficients in $\mathbb{Q}$ or in the field $\mathbb{R} \cap \overline{\mathbb{Q}}$ of real algebraic numbers, so that the polynomial admits a finite encoding suitable for a Turing machine). In fact, Tarski (1951) gave a decision procedure, based on elimination of quantifiers for $\mathbb{R}$ in the language of ordered fields, for the following more general problem.

Decision problem for the first-order theory of the ordered field $\mathbb{R}$

input: a first-order sentence ϕ in the language of ordered fields
question: Is ϕ true when the variables run over elements of $\mathbb{R}$?

9.1.2 Adjoining the exponential function

If one tries to extend this by allowing expressions involving also the real exponential function, then one runs into questions of transcendental number theory whose answer is still unknown. For example, can one decide for which rational numbers r, s, t the equation

$$e^{e^r} + e^s + t = 0$$

holds? But assuming Schanuel's conjecture (Lang 1966, pp. 30–31), which rules out such "accidental identities," Macintyre and Wilkie (1996) gave a decision algorithm for all first-order sentences for R with exponentiation in addition to the usual operations and $\leq$.

Remark 9.1 In contrast, for the set $\mathcal{E}$ of *complex* functions built up from integers and z using addition, multiplication, and composing with e^z, Adler

(1969, Theorem 1) proved that it is impossible to decide whether a finite list of functions in $\mathcal{E}$ has a common zero in $\mathbb{C}$. This can be proved by reduction to Hilbert's tenth problem, using two observations.

1. One can characterize $\mathbb{Q}$ in $\mathbb{C}$ as the set of ratios of zeros of $e^z - 1$.
2. One can characterize $\mathbb{Z}$ as the set of $x \in \mathbb{Q}$ such that there exists $z \in \mathbb{C}$ with $e^z = 2$ and $e^{zx} \in \mathbb{Q}$.

9.1.3   Adjoining the sine function

Adjoining most other transcendental functions leads quickly to undecidable problems. For example, consider the following, a variant of a theorem of Richardson.

Theorem 9.2 (cf. Richardson 1968, §1, Corollary to Theorem One) *There is a polynomial $P \in \mathbb{Z}[t, x_1, \ldots, x_n, y_1, \ldots, y_n]$ such that for each $a \in \mathbb{Z}$, the real analytic function*

$$P(a, x_1, \ldots, x_n, \sin \pi x_1, \ldots, \sin \pi x_n)$$

on $\mathbb{R}^n$ is either everywhere greater than 1, or else assumes values less than -1 and values greater than 1, but it is impossible to decide which, given a.

Sketch of proof. By Theorem 8.1, we can find a polynomial $p(t, \vec{x}) \in \mathbb{Z}[t, x_1, \ldots, x_n]$ defining a diophantine subset A of $\mathbb{Z}$ that is not computable. A little analysis shows that there is another polynomial $G \in \mathbb{Z}[t, x_1, \ldots, x_n]$ whose values are positive and growing so quickly that if we define

$$L_a(\vec{x}) := -2 + 4p(a, \vec{x})^2 + G(a, \vec{x}) \sum_{i=1}^{n} \sin^2 \pi x_i,$$

then $L_a(\vec{x}) \leq 1$ holds only in tiny neighborhoods of the integer solutions to $p(a, \vec{x}) = 0$. If $a \in A$, then such integer solutions exist and L_a takes the value -2 at those integer solutions and large positive values at some points with half-integer coordinates; otherwise, $L_a(\vec{x}) > 1$ on $\mathbb{R}^n$.

Remark 9.3 Richardson's original statement and proof of Theorem 9.2 were slightly more involved because they came before Hilbert's tenth problem had been proved undecidable. Richardson instead had to use the undecidability result for exponential diophantine equations mentioned in Section 8.1.

Laczkovich (2003) found a variant of Theorem 9.2 letting one use $\sin x_i$ in place of $\sin \pi x_i$ for all i. Also, there exist functions $h \colon \mathbb{R} \to \mathbb{R}^n$ with dense image, such as

$$h(x) := \left(x \sin x, x \sin x^3, \ldots, x \sin x^{2n-1} \right),$$

(this function, used by Denef and Lipshitz in (1989, Lemma 3.2), is a simpler version of one used by Richardson (1968, §1, Theorem Two)). By composing a multivariable function with h, one obtains analogues of Theorem 9.2 for functions of *one* variable:

Theorem 9.4 (cf. Laczkovich (2003), Theorem 1, which improves upon Richardson (1968), Corollary to Theorem Two) *Let $\mathcal{S}$ be the set of functions $\mathbb{R} \to \mathbb{R}$ built up from integers and x using addition, multiplication, and composing with* sin. *Then it is impossible to decide, given $f \in \mathcal{S}$, whether f is everywhere nonnegative. Deciding whether f is everywhere positive or whether f has a zero is impossible too.*

For later use, we record the fact that there exist functions $F_a \in \mathcal{S}$, depending in a computable way on an integer parameter a, such that either $F_a(x) > 1$ on $\mathbb{R}$, or else $F_a(x)$ assumes values less than -1 and values greater than 1, but it is impossible to decide which.

9.2 Equality of functions

Automatic homework graders sometimes need to decide whether two expressions define the same function. But deciding whether $|f(\vec{x})|$ is the same function as $f(\vec{x})$ amounts to deciding whether $f(\vec{x})$ is everywhere nonnegative, which, by Section 9.1.3, is impossible for $f \in \mathbb{Z}[x_1, \ldots, x_n, \sin x_1, \ldots, \sin x_n]$ or for $f \in \mathcal{S}$ (cf. Richardson 1968, §2, Theorem Two).

For further undecidability results in analysis deduced from the negative answer to Hilbert's tenth problem, see Adler (1969), Denef and Lipshitz (1989) and Stallworth and Roush (1997).

9.3 Integration

There exists an entire function on $\mathbb{C}$ whose derivative is e^{z^2}. But work of Liouville shows that no such function is expressible by an elementary formula, in the following sense.

For a connected open subset U of $\mathbb{C}$, let $\mathcal{M}(U)$ be the field of meromorphic functions on U. Say that a function $g \in \mathcal{M}(U)$ is elementary if it belongs to the last field K_n in a tower $\mathbb{C}(z) = K_0 \subset K_1 \subset \cdots \subset K_n$

of subfields of $\mathcal{M}(U)$ such that each extension K_{i+1} over K_i is either algebraic or obtained by adjoining to K_i either e^f or a branch of $\log f$ defined on U for some $f \in K_i$. For instance, the trigonometric functions and their inverses on a suitable U are elementary functions.

Liouville (1835, §VII) proved a general theorem that implies that there is no elementary antiderivative of e^{z^2} on any U. (Earlier, Liouville (1833) proved that certain algebraic functions, such as $(1 + x^4)^{-1/2}$, have no elementary antiderivative.) See Rosenlicht (1972) for an account of Liouville's methods.

Can one decide whether a given elementary function has an elementary antiderivative? Building on the work of Liouville, Risch (1970) sketched a positive answer to a precise version of this question. To obtain a positive answer, the question must be formulated carefully to avoid having to answer questions about whether a constant or function is identically 0. For example, it is not clear whether we can decide, given rational numbers r, s, t, whether

$$\int (e^{e^r} + e^s + t) e^{x^2} \, dx$$

is an elementary function.

Risch avoids this difficulty by restricting attention to functions in a tower of fields in which the constant field is an algebraically closed field of characteristic 0 with a specified finite transcendence basis, and in which each successive extension in the tower is either an explicit algebraic extension or an extension adjoining $\exp f$ or a branch of $\log f$ that does not change the field of constants.

Remark 9.5 If we try to generalize by allowing expressions involving the absolute value function $|\ |$, we encounter undecidability, as we now explain (cf. Richardson 1968, §2, Theorem Three). Define

$$\sigma(x) := \frac{1}{2}(|x| - |x - 1| + 1) = \begin{cases} 0, & \text{if } x \leq 0 \\ x, & \text{if } 0 < x < 1 \\ 1, & \text{if } x \geq 1. \end{cases} \qquad (2)$$

Recall the functions F_a at the end of Section 9.1.3. Then $\sigma(-F_a(x))$ is either 0 on all of $\mathbb{R}$, or it agrees with 1 on some open interval, but we cannot decide which. Thus we cannot decide whether

$$\int \sigma(-F_a(x)) e^{x^2} \, dx$$

is an elementary function on all of $\mathbb{R}$.

Remark 9.6 Deciding whether an improper integral converges is undecidable too, as was observed by P. S. Wang (1974). Specifically, we cannot decide whether

$$\int_{-\infty}^{\infty} \frac{1}{(x^2 + 1)F_a(x)^2} \, dx$$

converges.

9.4 Differential equations

Consider algebraic differential equations (ADEs)

$$P\left(x, y, y', y'', \ldots, y^{(n)}\right) = 0$$

to be solved by a function y of x, where P is a polynomial with integer coefficients. Denef and Lipshitz (1989, Theorem 4.1) proved that the following problem is undecidable.

Existence of solutions to algebraic differential equations
  input: $P \in \mathbb{Z}[x, y_1, y_2, \ldots, y_n]$
  question: Does $P(x, y, y', y'', \ldots, y^{(n)}) = 0$ admit a real analytic solution on $[0, \infty)$?

It remains undecidable even if one restricts the input so as to allow only ADEs that have a unique analytic solution in a neighborhood of 0. The idea of the proof is to consider a function built up using sin as in Section 9.1.3 for which one cannot decide whether it is either everywhere positive, and then to show that its reciprocal satisfies an ADE.

Remark 9.7 To avoid having to use π in the coefficients of P, Denef and Lipshitz observed that $\tan^{-1} x$ is a function satisfying an ADE such that $\lim_{x \to +\infty} \tan^{-1} x = \pi/2$. An alternative would be to use the approach of Laczkovich (2003) for eliminating π.

Remark 9.8 ADEs can behave strangely in other ways too. L. Rubel (1981) constructed a single explicit ADE whose solutions approximate any continuous function: more precisely, for any continuous functions $f \colon \mathbb{R} \to \mathbb{R}$ and $\varepsilon \colon \mathbb{R} \to \mathbb{R}_{>0}$, there exists a C^∞ solution $g \colon \mathbb{R} \to \mathbb{R}$ to the ADE satisfying $|g(x) - f(x)| < \varepsilon(x)$ for all $x \in \mathbb{R}$.

For other results and questions concerning existence and computability of solutions to differential equations, see Jaśkowski (1954), Adler (1969),

Aberth (1971), Pour-El and Richards (1979, 1983), Rubel (1983, 1992) and Denef and Lipshitz (1984).

10   Dynamical systems

Many nonlinear dynamical systems are capable of simulating universal Turing machines, and hence they provide undecidable problems.

10.1   Dynamical systems on $\mathbb{R}^n$

Call a map $\mathbb{R}^n \to \mathbb{R}^m$ affine linear if it is a linear map plus a constant vector. Call a map $f : \mathbb{R}^n \to \mathbb{R}^m$ piecewise affine linear if $\mathbb{R}^n$ can be partitioned into finitely many subsets U_i, each defined by a finite number of affine linear inequalities such that $f|_{U_i}$ agrees with an affine linear map depending on i. Call such a map rational if all the coefficients of the affine linear polynomials involved are rational. Given such a map f, let f^k be its kth iterate. Moore (1990) proved that the following problem is undecidable.

Point goes to origin in finite time

> **input:** a rational piecewise affine linear map $f : \mathbb{R}^2 \to \mathbb{R}^2$ and a point $\vec{a} \in \mathbb{Q}^2$
>
> **question:** Does there exist k such that $f^k\left(\vec{a}\right) = \vec{0}$?

Similarly, Siegelmann and Sontag (1995) proved that neural nets can simulate a universal Turing machine: in particular, if $\sigma : \mathbb{R}^n \to \mathbb{R}^n$ is the map obtained by applying the function (2) coordinatewise, then there exists n and a specific matrix $A \in M_n(\mathbb{Z})$, for which it is impossible to decide, given a starting point $\vec{a} \in \mathbb{Q}^n$, whether some iterate of $\sigma(A \vec{x})$ maps $\vec{a}$ to $\vec{0}$.

Instead of asking about the trajectory of one point, one can ask global questions about the dynamical system, such as whether every trajectory converges. Blondel *et al.* (2001) prove that many such questions are undecidable for piecewise affine linear maps.

For further results relating dynamical systems and computability, see the survey article by Blondel and Tsitsiklis (2000).

10.2   Dynamical systems on the set of positive integers

There are also undecidable problems concerning dynamics of maps $f : \mathbb{Z}_{>0} \to \mathbb{Z}_{>0}$ such as

$$f(x) := \begin{cases} 3x + 1, & \text{if } x \text{ is odd} \\ x/2, & \text{if } x \text{ is even.} \end{cases}$$

The Collatz $3x + 1$ problem, which asks whether for every $n \in \mathbb{Z}_{>0}$, there exists k such that $f^k(n) = 1$, has been open since the 1930s (Lagarias 1985); see Lagarias (2010) for a recent compilation of articles on this subject. The following generalization has been proved undecidable.

Generalized Collatz problem

> **input:** $m \in \mathbb{Z}_{>0}, a_0, \ldots, a_{m-1}, b_0, \ldots, b_{m-1} \in \mathbb{Q}$ such that the function f given by $f(x) = a_i x + b_i$ for $x \bmod m = i$ maps $\mathbb{Z}_{>0}$ to itself
>
> **question:** Is it true that for every $n \in \mathbb{Z}_{>0}$ there exists k such that $f^k(n) = 1$?

For this and related results, see the papers by Conway (1972), Kurtz and Simon (2007), and Endrullis *et al.* (2009).

11 Probability

Consider a random walk on the set $(\mathbb{Z}_{\geq 0})^n$ of lattice points in the non-negative orthant. At each time, the walker takes a step by adding a vector in $\{-1, 0, 1\}^n$. If $\vec{x} = (x_1, \ldots, x_n)$ is the current position, the vector to add is chosen with respect to a probability distribution Λ_S depending only on the set $S = \{i : x_i \neq 0\}$, and Λ_S is such that the walker never leaves the orthant. Suppose also that every probability in the description of each Λ_S is in $\mathbb{Q}$. Say that the random walk starting at $\vec{a}$ is stable if there exists $C > 0$ such that with probability 1 the walker returns to $\{\vec{x} : |\vec{x}| < C\}$ infinitely often.

Stability of random walks

> **input:** $n \in \mathbb{Z}_{\geq 0}$, probability distributions Λ_S as above for $S \subseteq \{1, \ldots, n\}$, and $\vec{a} \in (\mathbb{Z}_{\geq 0})^n$
>
> **question:** Is the random walk starting at $\vec{a}$ stable?

Gamarnik (2002) proved that this problem is undecidable even if all the probabilities are 0 or 1! To do this, he showed that any Turing machine could be simulated by such a deterministic walk.

Moreover, many basic questions about the stationary distribution of a random walk as above turn out to be undecidable, even if one assumes that the stationary distribution exists (Gamarnik 2007).

12 Algebraic geometry

12.1 Rational sections

Kim and Roush (1992) proved the undecidability of Hilbert's tenth problem for the field $\mathbb{C}(t_1, t_2)$ of rational functions in two variables. (Strictly speaking, one should assume that the input has coefficients in $\overline{\mathbb{Q}}(t_1, t_2)$ instead of $\mathbb{C}(t_1, t_2)$, for the sake of encoding it for input into a Turing machine, but we will ignore this subtlety from now on.) By the same argument as in Section 8.2.1, Hilbert's tenth problem for $\mathbb{C}(t_1, t_2)$ is equivalent to the problem of deciding whether a $\mathbb{C}(t_1, t_2)$-variety has a $\mathbb{C}(t_1, t_2)$-point (any pair of equations $f = g = 0$ can be converted to $f^2 + t_1 g^2 = 0$). Eisenträger (2004), using work of Moret-Bailly (2005), generalized the Kim–Roush result to the function field of any fixed irreducible $\mathbb{C}$-variety S of dimension at least 2. Whether Hilbert's tenth problem for $\mathbb{C}(t)$ is undecidable is an open question, studied in Kollár (2008). It is also open for the function field of each other curve over $\mathbb{C}$.

Let us return to the Kim–Roush result. By interpreting the "constants" t_1 and t_2 as variables, one can associate to each $\mathbb{C}(t_1, t_2)$-variety X a $\mathbb{C}$-variety Y equipped with a rational map $\pi : Y \to \mathbb{A}^2$. The $\mathbb{C}(t_1, t_2)$-points of X correspond to rational sections of π, i.e., to rational maps $s \colon \mathbb{A}^2 \to Y$ such that $\pi \circ s$ is the identity. This dictionary translates the Kim–Roush result into the undecidability of the following problem.

Existence of rational sections
  input: a $\mathbb{C}$-variety Y and a rational map $\pi : Y \to \mathbb{A}^2$
  question: Does π admit a rational section?

12.2 Automorphisms

Using the undecidability of Hilbert's tenth problem, one can show that it is impossible to decide, given a variety X, a point $x \in X$, and a subvariety $Z \subset X$, whether there exists an automorphism of X mapping x into Z (Poonen 2011). In fact, there are fixed X and x for which the problem for a variable input Z is undecidable. More precisely, there is a smooth projective geometrically irreducible $\mathbb{Q}$-variety X and a point $x \in X(\mathbb{Q})$ such that the following problem is undecidable.

Automorphisms mapping a point into a subvariety
  input: a smooth projective geometrically irreducible subvariety $Z \subset X$
  question: Does there exist an automorphism of X mapping x into Z?

Moreover, X can be chosen so that all its automorphisms over any field extension are already defined over $\mathbb{Q}$, so it does not matter whether we require the automorphisms to be defined over the base field.

On the other hand, the following question has remained open.

Question 12.1 Is there an algorithm to decide whether a given variety has a nontrivial automorphism?

Possibly related to this is the following.

Question 12.2 Given an f.p. group G, can one effectively construct a variety X_G whose automorphism group is G?

A positive answer to Question 12.2 would yield a negative answer to Question 12.1, since it is impossible to decide whether an f.p. group is trivial (Corollary 6.3).

12.3 Isomorphism

Given the undecidability of the homeomorphism problem for manifolds, it is natural to ask for the algebraic geometry analogue:

Variety isomorphism problem
> **input:** two varieties X and Y over $\overline{\mathbb{Q}}$
> **question:** Is $X \simeq Y$?

The question of whether this problem might be undecidable was asked of the author by B. Totaro in 2007.

We stated the problem over $\overline{\mathbb{Q}}$, because most algebraic geometry is done over an algebraically closed field and we wanted the input to admit a finite description. Alternatively, we could work over an algebraically closed field $\overline{\mathbb{Q}(t_1, t_2, \ldots)}$ of countable transcendence degree over $\mathbb{Q}$. This would capture the essence of the problem over $\mathbb{C}$, since any pair of varieties over $\mathbb{C}$ may be simultaneously defined over a finitely generated subfield of $\mathbb{C}$ and the existence of an isomorphism is unaffected by enlarging the ground field from one algebraically closed field to another. One could also consider other fields, such as $\mathbb{Q}, \overline{\mathbb{F}}_p, \mathbb{F}_p$ or $\overline{\mathbb{F}_p(t_1, t_2, \ldots)}$. There is no field over which it is known whether one can solve the variety isomorphism problem.

12.4 Birational equivalence

It is also unknown whether there is an algorithm to decide whether two given varieties are birationally equivalent. On the other hand, given an explicit rational map, one can decide whether it is a birational map, and whether it is an isomorphism.

Remark 12.3 For algebraic geometers: if at least one of the varieties X and Y over $\overline{\mathbb{Q}}$ is of general type, then the set of birational maps $X \to Y$ is finite and computable (see below), and we can decide which of these birational maps are isomorphisms, and hence solve the variety isomorphism problem in this restricted setting. Matsumura (1963) proved that the birational automorphism group of a variety of general type is finite. The set of birational maps $X \to Y$ is either empty or a principal homogeneous space under this group, so it is finite too. Let us sketch an algorithm for computing this set. For $n = 1, 2, \ldots$, compute the maps determined by the pluricanonical linear system $|nK|$ for X and Y until an n is found for which at least one of the two maps is birational onto its image. Then the other must be too, and the linear systems must have the same dimension, say n, since otherwise we know already that $X \simeq Y$. The birational maps are then in bijection with the linear automorphisms of $\mathbb{P}^n$ mapping one canonical image to the other, and we can find equations for the locus of these automorphisms as a (finite) subscheme of PGL_{n+1}.

13 Algebra

13.1 Commutative algebra

If we restrict the variety isomorphism problem to the category of *affine* varieties, we obtain, equivalently, the isomorphism problem for finitely generated $\overline{\mathbb{Q}}$-algebras. Here each algebra can be presented concretely as $\mathbb{Q}[x_1, \ldots, x_n]/(f_1, \ldots, f_m)$ by specifying n and explicit polynomials $f_1, \ldots, f_m$.

One can replace $\mathbb{Q}$ by other rings of constants (whose elements can be encoded for computer input). For example, taking $\mathbb{Z}$ yields the following problem.

Commutative ring isomorphism problem
> **input:** two finitely generated commutative rings A and B
> **question:** Is $A \simeq B$?

The undecidability status of this problem is unknown. In fact, the status is unknown also for the isomorphism problem for finitely generated commutative algebras over any fixed nonzero commutative ring (with elements encoded such that addition and multiplication are computable).

13.2 Noncommutative algebra

The noncommutative analogue of the previous problem is undecidable, as we now explain. Let $\mathbb{Z}\langle x_1, \ldots, x_n \rangle$ be the noncommutative polynomial ring (free associative algebra with 1) in n variables over $\mathbb{Z}$. A (possibly noncommutative) f.p. $\mathbb{Z}$-algebra is the quotient of $\mathbb{Z}\langle x_1, \ldots, x_n \rangle$ by the 2-sided ideal generated by a finite list of elements $f_1, \ldots, f_m$. For an f.p. group G, the group ring $\mathbb{Z}G$ is an f.p. $\mathbb{Z}$-algebra, and $\mathbb{Z}G \simeq \mathbb{Z}$ if and only if $G \simeq \{1\}$. So if there were an algorithm to decide whether two f.p. $\mathbb{Z}$-algebras were isomorphic, we could use it to decide whether an f.p. group is trivial, contradicting Corollary 6.3.

For other undecidable problems concerning noncommutative f.p. algebras, see Anick (1985).

14 Games

14.1 Abstract games

Given $m \geq 1$ and a computable function $W : \mathbb{N}^m \to \{A,B\}$, consider the two-player game of no chance in which

- the players (A and B) alternately choose natural numbers, starting with A, and ending after m numbers $x_1, \ldots, x_m$ have been chosen;
- there is perfect information (both players know the rules and can see all previously made choices); and
- the winner is $W(x_1, x_2, \ldots, x_m)$.

Many games can be fit into this framework.

A result of Kalmár (1928–1929), building on work of Zermelo (1913) and König (1927), states that exactly one of the two players has a winning strategy. But we have the following.

Theorem 14.1 *It is impossible to decide, given m and W, which player has a winning strategy.*

Proof. Given a program p, consider the one-move game in which A chooses a positive integer x_1 and wins if p halts within the first x_1 steps. Player A has a winning strategy if and only if p halts, which is undecidable.

More surprising is the following result of Rabin (1957).

Theorem 14.2 *There is a three-move game in which B has a winning strategy, but not a computable winning strategy (i.e., there is no computable function of x_1 that is a winning move x_2 for B).*

Proof. Post (1944, §5) proved that there exists a simple set, i.e., a c.e. set $S \subset \mathbb{N}$ whose complement $\overline{S}$ is infinite but contains no infinite c.e. set. Fix such an S. Let $g : \mathbb{N} \to \mathbb{N}$ be a computable function with $g(\mathbb{N}) = S$. Consider the three-move game in which A wins if and only if $x_1 + x_2 = g(x_3)$.

Player B's winning strategy is to find $t \in \overline{S}$ with $t > x_1$, and to choose $x_2 = t - x_1$. A computable winning strategy $x_2 = w(x_1)$, however, would yield an infinite c.e. subset $\{x_1 + w(x_1) : x_1 \in \mathbb{N}\}$ of $\overline{S}$.

Remark 14.3 Using the undecidability of Hilbert's tenth problem, Jones (1982) gave new proofs of these theorems using games in which W simply evaluates a given polynomial at the m-tuple of choices to decide who wins.

Hearn (2006) proved that team games with imperfect information can be undecidable even if they have only *finitely* many positions! For an account of this work and a complexity analysis of many finite games, see Hearn and Demaine (2009).

14.2   Chess

Stanley (2010) asked whether the following problem is decidable.

Infinite chess
  input: A finite list of chess pieces and their starting positions on a
  $\mathbb{Z} \times \mathbb{Z}$ chessboard
  question: Can White force mate?

For a precise specification of the rules, and for related problems, see Brumleve *et al.* (2012).

It is unknown whether this problem is decidable. On the other hand, Brumleve *et al.* (2012) showed that one *can* decide whether White can mate in n moves if a starting configuration and n are given. This statement can be proved quickly by encoding each instance of the problem as a first-order sentence in Presburger arithmetic, which is the theory of $(\mathbb{N}; 0, 1, +)$. (Presburger arithmetic, unlike the theory of $(\mathbb{N}; 0, 1, +, \cdot)$, is decidable (Presburger 1929).)

15   Final remarks

Each undecidable problem P we have presented is at least as hard as the halting problem H, because the undecidability proof ultimately depended on encoding an arbitrary instance of H as an instance of P. In the other direction, many of these problems could be solved if one could decide whether a certain search terminates; for these P, an arbitrary instance of P can be encoded as an instance of H. The problems for which both reductions are possible are called c.e. complete, and they are all of exactly the same difficulty with respect to Turing reducibility. For example, an algorithm for deciding whether a finitely presented group is trivial could be used to decide whether multivariable polynomial equations have integer solutions, and vice versa!

On the other hand, certain other natural problems are strictly harder than the halting problem. One such problem is the generalized Collatz problem of Section 10.2; see Kurtz and Simon (2007) and Endrullis *et al.* (2009).

CHAPTER 11

Reflecting on logical dreams

Saharon Shelah

1 The independence phenomenon

The "forcing method," introduced by Cohen (1966), has produced a wealth of independence results in set theory. What about PA, that is, arithmetic? Now Shelah (2003a) described the following Dream 2.1:

> Find a "forcing method" relative to PA which shows that PA and even ZFC does not decide "reasonable" arithmetical statements, in analogy with the situation in which the known forcing method works for showing that ZFC cannot decide reasonable set theoretic questions; even showing the unprovability of various statements in bounded arithmetic (instead of PA) is formidable.

Why is it interesting to prove independence results? I can understand someone disregarding the work on cardinal arithmetic, claiming that it is not interesting, though I think he or she is wrong. If I had to act as a lawyer I could try to write an argument for him or her; I could express myself as "(s)he is wrong but consistent." But in this case it is hard for me to understand any opposition (well, from mathematicians, let us say pure mathematicians).

A mathematician can look at the work of Gödel and Cohen and say "very nice, but this does not really affect me." In fact, this was the outlook of not a few mathematicians in various fields before forcing was proved relevant to their work. A finite combinatorialist can look at all the independence results and say that they do not deal with problems he or any reasonable mathematician has been working on.

We try here to comment on Shelah (2003a), give clarifications, and also discuss what we have learnt since, finishing with a recapitulation of the main points (in Section 8). This was originally written as answers to questions by Juliette Kennedy. We thank Gregory Cherlin, Udi Hrushovski, Juliette Kennedy, Menachem Kojman and Jouko Väänänen for helpful comments and questions and Kennedy and Väänänen again for editorial assistance. This is paper E73 in the author's list of publications.

On pointing out that this can be translated to problems about the existence of solutions to polynomials (for Gödel, a polynomial with exponentiation, after J. Robinson–Putnam–Matiyasevich, a real polynomial) he or she will probably raise his or her head and say: "in principle it may be interesting, though I do not understand the hubris of Hilbert who thought to find an algorithm, but we are working on quite specific problems. In fact, nothing close to Ackermann functions (not to say faster growing functions) interests me; so, on the one hand the Paris–Harrington theorem and its descendants and relatives are irrelevant and, on the other hand, the polynomials mentioned above are artificial. It is very nice to know that the earth is not flat, but if I do not know anything about places far from me and if my technology is primitive, this has no real affect on me."

The dream above should change this.

2   Semi-axioms

The independence results of set theory raise the question, of what axioms could we add to the ZFC axioms in order to decide some or most of the statements left undecided by ZFC. It is a difficult question, and famously discussed by Gödel, what the criteria for new axioms should be. We have talked in this connection about "semi-axioms":

> What are our [criteria] for semi-axioms? First and most important, a semi-axiom must have many consequences, making it have a rich, deep, beautiful theory. Second, it is preferable that a semi-axiom is reasonable and "has positive measure." (Shelah 2003a, p. 214)

There is a striking difference between the first and the second criteria. Let us try to consider examples close to my heart. The axiom $V = L$ is an excellent example for the first criterion; there are many consequences in the works of Gödel, Jensen and others. But $V = L$ is not seen as being reasonable and having "positive measure," at least in the opinion of most set theorists.

Another example is the axiom "the continuum is real valued measurable," which Fremlin wrote so much about. Well, by the first criterion it is reasonably strong (though nothing like $V = L$) but its measure, it seemed, is not impressive. (See also, for example the universe forced in Blass and Shelah (1987).) On the other hand "the continuum is weakly inaccessible" seems to be very reasonable but has few consequences.

The so-called "California school" advocates the importance of determinacy axioms in descriptive set theory, and the relevance of such axioms being consequences of certain large cardinal axioms, and moreover, the importance of their being true. Is this a reasonable position? In Shelah (2003a) I made the following remark:

> I strongly reject the California school's position on several grounds.
>
> (a) Generally I do not think that the fact that a statement solves every-thing really nicely, even deeply, even being the best semi-axioms (if there is such a thing, which I doubt) is a sufficient reason to say it is a "true axiom." In particular I do not find it compelling at all to see it as true.
>
> (b) The judgments of certain semi-axioms as best is based on the groups of problems you are interested in. For the California school descrip-tive set theory problems are central. While I agree that they are important and worth investigating, for me they are not "the center." Other groups of problems suggest different semi-axioms as best, other universes may be the nicest from a different perspective.
>
> (c) Even for descriptive set theory the adoption of the axioms they advocate is problematic. It makes many interesting distinctions disappear.

I still stand by (a), (b), and (c) as I stated them above.

For (b), we can consider the following two examples. For descriptive set theory, the axiom $V = L$ gives very definitive answers, still "the hierarchy collapses" and descriptive set theorists are not satisfied. Also the axiom AD gives very definitive answers, but ones which seem more natural, and are much more liked by descriptive set theorist. Still the knowledge that under $V = L$ we get other answers is illuminating.

Note that Woodin, on learning more, changed his mind on the value of the continuum, but I do not know enough to comment on this.

As to (a), note that we may well believe that such a theory ZFC + X is natural, important, elegant and will even appear in unexpected contexts *but* we may also know that another semi-axiom Y contradicting it has all these wonderful properties too.

Also, the reader may like to refer to the discussion in the sixth paragraph of Section 3.2 in Shelah (2003a). With this in mind, let us return, as an example, to another family of problems for which $V = L$ gives a definite answer. The problems are on Abelian groups and the answers erased distinctions. As Nunke (1977) explains nicely, the original solution of Whitehead's problem is not so satisfying. The problem as stated was whether every Whitehead group is free, where G is a Whitehead group, when it is Abelian and if H is an Abelian group extending Z with the quotient being isomorphic to G then Z is a direct summand. The solution was that if $V = L$ then every Whitehead group is free (as an Abelian group) while if MA (the Martin axiom) is true and CH fails this is not the case. Now Nunke (1977) says that really there is a family of related properties (e.g., 'hereditary separable) and the real problem is to sort them out; and also to find whether "$\mathbb{Q} = \text{Ext}(G, Z)$ for some G." Indeed this was done, see Eklof and Mekler (2002). If $\mathbf{V} = \mathbf{L}$ we get that the related classes are mostly equal etc., and under some forcing axiom the answer is that they are distinct. But we may wonder whether there is an axiom which gives a different coherent picture, antithetical to the one in $\mathbf{L}$. This is the subject and aim of Shelah (undated d) in which we force a universe $\mathbf{V}$ in which GCH holds, but for every regular uncountable λ and for every suitable uniformization property, it holds for some suitable stationary subset of λ, say with a pregiven cofinality. This answers many problems, for example in Abelian group theory, "nicely." Properties are not equivalent without good reasons. So we may look at this axiom as trying to do for this family of Abelian group problems (and hopefully more) what $\text{AD}_{L[\mathbb{R}]}$ does for descriptive set theory problems, though it is far from being as grand as it is.

In another direction, in some model theoretic work of the author, it appears that assuming instances of WGCH (the weak generalized continuum hypothesis, i.e., $2^\lambda < 2^{\lambda^+}$) is very helpful.[1] This motivated Baldwin (2009) to suggest that we should consider the axiom WGCH.

Concerning a large continuum, we may be even bolder.

Dream 2.1 Find a reasonable semi-axiom which not only implies that, say, many of the established cardinal invariants of the continuum are (pairwise) distinct but determines their order. For definiteness we may concentrate on Cichoń's diagram.

3 Large continuum

In Gödel's universe of constructible sets we have $2^{\aleph_0} = \aleph_1$. Remarkably, several different constructions in set theory seem to deliver $2^{\aleph_0} = \aleph_2$, most notably Martin's maximuum MM. Is this because we know how to iterate forcings only on this level, or is there some deeper reason? In Shelah (2003a), I formulated the following Dream (no. 3.4) on a large continuum:

> Develop a theory of iterated forcing for a large continuum as versatile as the one we have for $2^{\aleph_0} = \aleph_2$, and to a lesser extent, for $2^{\aleph_0} = \aleph_1$. (Shelah 2003a, Dream 3.4)

[1] See Baldwin (2009) and Shelah (2009a, 2009b).

Dream 4.5 says more or less the same thing from another perspective:

(a) *Find a real significance for $2^{\aleph_0} = \aleph_{753}$,*
(b) *or for $2^{\aleph_0} = \aleph_{\omega^3} + \omega + 5$,*
(c) *show that all values of $2^{\aleph_0}$ which are $> \aleph_2$ are similar in some sense (or at least all values $\aleph_n > \aleph_2$, all regular $\aleph_\alpha > \aleph_{\omega_1}$ or whatever).* (Shelah 2003a, Dream 4.5)

Note that in clause (a) above we mean to ignore the too easy example (if $f: [\aleph_n]^{n+1} \to [\aleph_n]^{\aleph_0}$ then there is an *n*-free set with $n + 2$ members).

Recently, Neeman has made a major advance using a new iteration. Clearly this gives the consistency of a generalization of the PFA (Proper Forcing Axiom) for large continuum (replacing $\aleph_1$ by a larger cardinal), but I am not familiar enough with it to say more about it.

Woodin, using his PMAX method, got the consistency of ZFC + maximal set of statements of the form "for every subset of $\aleph_1$ there is a subset of $\aleph_1$ such that . . . " It was natural to ask about replacing ZFC by ZFC + some axiom, see Shelah and Zapletal (1999). But for probably the most natural statement, CH, the answer has long eluded us.[2]

Recently this was solved. Aspero *et al.* (2013) proved that there exist sentences ψ_1 and ψ_2 which are Π_2 over the structure $(\mathcal{H}(\aleph_2), \in, \aleph_1)$ such that

1. ψ_2 can be forced by a proper forcing which does not add new reals;
2. if there exists a strongly inaccessible limit of measurable cardinals, then ψ_1 can be forced by a proper forcing which does not add new reals;
3. the conjunction of ψ_1 and ψ_2 implies that $2^{\aleph_0} = 2^{\aleph_1}$.

Even more recently Moore (2013) obtained remarkably more concrete results, for example, if CH then there is a tree with $\aleph_1$ nodes and $\aleph_1$ levels, which is a proper forcing adding no new reals, but forcing with it renders it not proper.[3] Weak progress was made in Shelah (2010, 2011a) on developing iterated forcing with large continuum to obtain results about cardinal invariants of the continuum, especially in the case that we want to force several invariants to have given values simultaneously. A different direction, so-called creature iteration, described by Kellner and Shelah (2009) is ψ. Somewhat better still in the direction of creature iteration, but still far from a real answer, is work in progress by Goldstern and Kellner on the consistency of distinguishing five cardinal invariants from each other in the

---

[2] Some relevant information can be found in Shelah (undated c).
[3] For complementary consistency see Shelah (2003b).

Cichoń diagram. Concerning this, more work using finite support iteration of c.c.c. forcing is being done by Brendle and coworkers.

4 Doing without the Axiom of Choice

As I said in Shelah (2003a), I am not fond of set theory with weak forms of choice and was not bothered by what have been considered unintuitive consequences of this, such as the so-called "paradoxical decomposition of the sphere" by Banach and Tarski (1924). Naturally, the Axiom of Choice, AC, is part of the standard axioms ZFC of set theory. Still, I proposed the following:

> *Develop combinatorial set theory for universes with limited amount of choice (see Shelah 2000d).* (Shelah 2003a, problem 6.16)

Clearly, AC is true. Set theory without AC has been considered as a prototypical example of a dead direction. Why should we consider this direction? There are several good reasons (in addition to a fascination with the resurrection of the dead). First, let us have intellectual honesty – failure to have a nice set theory contributed to the acceptance of AC, so if we also have a nice set theory without AC we should check this premise ("nice" means with interesting content). Second, a good motivation for proving results without AC is that it is quite reasonable to consider when we can have a definable solution to a problem, rather than a mere existence theorem. Third, and finally, I think ideology and/or good taste should not stop you from proving a good theorem.

It has long been known to me that there is no serious model theory without choice – none of the basic theorems (compactness, downward and upward Löwenheim–Skolem theorems) hold. Of course, using $L[X]$ some restricted versions hold, for example, compactness for first-order theories T with a well-orderable vocabulary. This position seems consistent with considering theorems and their proofs to see how much choice was used in the "input," for example, we can ask whether the theory we study is well ordered, or how much choice was used in the proof.

Some years ago, having considered a weak version of the Axiom of Choice (Ax_4, see below), we tried to see what happens to Morley's theorem (if T is a countable theory which is categorical in one cardinal $\aleph_\alpha > \aleph_0$, then T is categorical in every $\aleph_\alpha > \aleph_0$). On checking carefully, Morley's proof gives "there is a set of reals of cardinality $\aleph_1$." If we work further and quote later results in stability, we can prove this theorem in ZF, see Shelah (2009c). Considering the place of Morley's theorem in model theory this seems quite conclusive evidence against the thesis above.

What about the Main Gap (Shelah 1990, Chapter. XII)? This seems much harder. Probably in "$\dot{I}\,(\lambda,\ T){=}2^{\lambda}$" we should revise "$2^{\lambda}$".[4] What about powers which are not cardinals? When studying them, it is natural to consider *reasonable* powers: $|A|$ is called *reasonable* if A can be linearly ordered and $|A| \cdot |A| = |A|$. We can show that every countable first-order theory has a model in every reasonable power (and some first-order sentence has a model of power $|A|$ if and only if $|A|$ is a reasonable power) so we may consider categoricity in reasonable powers, but the work in Shelah (2009c) has not been continued so far. Mendick and Truss (2003) have worked on model theory with little choice but in a very different direction.

What about set theory? On the one hand some parts of set theory are not affected by the absence of choice; inner model theory is not harmed (and lower bounds on consistency strength are still interesting in this context); descriptive set theory (and $L[\mathbb{R}]$ under AD), and working out the relations between various weaker versions of choice are also unaffected. But combinatorial set theory is strongly affected: ZF alone and even ZFC + DC seems hopeless (recall that even Poincaré accepted DC).

Naturally, having pcf theory with weak choice seems interesting to me, and work seems to indicate that this direction is not empty (Shelah 1997b).

Considerably more has been done. Some researches have tried to investigate set theory under ZFC + DC + Ax_4, see below. Others have tried to assume ZFC + DC + AC_X where X is $\mathcal{P}(\kappa)$ or essentially $\mathcal{P}(\mathcal{P}(\kappa))$, and investigate $^{\kappa}\lambda$, and even consider RGCH (discussed below). Note that after Shelah (1994b), a presentation of pcf was included in Holz *et al.* (1999) and Abraham and Magidor (2010). There are also several application, see for example Rinot (2007). I have also continued this work (e.g., Shelah 1993a, 1996, 1997a, 2000a, 2000e, 2007, undated f).

Gitik has many additional works on forcing, in particular proving the consistency of the failure of $(WH)_2$, (proving that there may be $\aleph_1$ cardinals of cofinality $\aleph_0$ whose pp is above their supremum). But the parallel major result of Shelah (1997b) remains open. The question is: can there be a strong limit singular λ of cofinality κ and increasing continuous sequence $\langle \mu_i : i < k \rangle$ with limit λ and stationary subsets S_1, S_2 of κ such that $(\prod_i \in S_l \mu_i^{+l}, < J_k^{bd})$ has true cofinality λ^{+l}?

Close to my heart is the RGCH, the revised generalized continuum hypothesis (see Shelah (2000b, 2002, 2006) which give more accessible proof). It asserts that we should refine the definition of power to $\lambda^{[\kappa]}$ such that it is not monotonic in κ (still $\lambda^{\kappa} = \lambda$ if and only if $(\forall \theta)\ \theta = \mathrm{cf}(\theta) \leq \kappa \to \lambda^{[\kappa]} = \lambda$).

---

[4] For still more on dichotomies related to "$\dot{I}\,(\aleph_\alpha,\ T) \geq |\alpha|$" see Shelah (1990).

Our reward is now that for every λ for most $\kappa \ll \lambda$ we have $\lambda^{[\kappa]} = \lambda$; recall that $\lambda^{[\kappa]} = \min\{|\mathcal{P}|$: every subset of λ of cardinality κ is the union of $<\kappa$ members of $\mathcal{P}\}$. More specifically, the theorem says that for every $\lambda \geq \beth_\omega$ for every large enough $\kappa < \beth_\omega$ we have $\lambda^{[\kappa]} = \lambda$. Kojman and Soukup have used this result, and Kojman has written on the history of singular cardinals.

Now in Shelah (2011b, undated j), an attempt is made to find parallels to pcf theorems under weak choice. The pcf theorem is generalized by replacing "cofinal sequence of members of $\Pi\mathfrak{a}$ modulo a filter" by "a sequence of subsets of the product which is increasing and cofinal." In Shelah (undated i) it is shown that $^\kappa\lambda$ can be decomposed to a few well-ordered sets, where few means not depending on λ, only on κ. Larson and Shelah (2009) showed that a theorem on partition of a stationary set into many stationary subsets can be generalized.

Let us return to Ax_4. Shelah (undated i) suggests a direction orthogonal to $\mathbf{L}[\mathbb{R}]$, recalling that in $\mathbf{L}[\mathbb{R}]$ the only reason for non-well-orderability is $\mathbb{R}$. The condition suggested is Ax_4: "$[\lambda]^{\aleph_0}$ is well orderable for every cardinal λ," and also weaker relatives. The thesis is that in this context "set theory is not so far from normal," the only reasons for non-well-orderability are the "$\mathcal{P}(\kappa)$ not well ordered" for regular κ or sometimes $\mathcal{P}(\mathcal{P}(\kappa))$. In what way does Shelah (undated i) show that ZFC + DC + Ax_4 is "nice"? For example, (provably in it) there is a proper class of regular successor cardinals, and $[\lambda]^\kappa$ can be almost well ordered, modulo $\mathcal{P}(\mathcal{P}(\kappa))$, see above. So still we can have successor cardinals which are singular, but not too many. In fact, it suggests the following.

Thesis ZF + DC + Ax_4 is an interesting set theory to investigate.

A relevant universe is the following variant of Easton's model: for regular $\kappa > \aleph_0$ we add many subsets but not a well ordering. We may try to investigate ZF + DC + Ax_4 more systematically. This is the aim of the forthcoming Shelah (undated h), which in particular gives an effective version of the pcf theorem for any set a of regular cardinals that are large enough ($\geq \theta$ which is about the size of $\mathrm{hrtg}(\mathcal{P}(\mathcal{P}(\mathfrak{a})))$, the Hartog number of $\mathcal{P}(\mathcal{P}(\mathfrak{a}))$). This restriction is a serious loss. But the assumption "$\mathfrak{a}$ is a set of *regular* cardinals" is not so natural in the present context. However, we can correct this: $\mathfrak{a}$ can be a set of just limit ordinals, possibly each of cofinality $\aleph_0$, but all is done modulo the ideal:

$$\mathrm{cf-id}_\theta(\mathfrak{a}) = \{\mathfrak{b} \subseteq \mathfrak{a} : \text{there is a set } u \subseteq \sup(\mathfrak{a}) \text{ of cardinality} < \theta$$
$$\text{such that } \forall \delta \in \mathfrak{b}(\delta = \sup(u \cap \delta))\}.$$

5 Model theory

The so-called Main Gap theorem is for countable first-order theories. In Shelah (2003a) the following proposal is made:

> Prove a form of the main gap for $\psi \in \mathbb{L}_{\lambda+,\aleph_0}$ (or just $\mathbb{L}_{\aleph_1,\aleph_0}$); i.e., for every such ψ either $\dot{I}(\lambda, \psi) > \lambda$ (where $\dot{I}(\lambda, \psi) = |\{M/\cong : M \models \psi, |M| = \lambda\}|$) for every λ large enough or there is an ordinal γ such that for every ordinal α, $\dot{I}(\aleph_\alpha, \psi) \leq \beth_\gamma (|\alpha|)$. (Shelah 2003a, Question 6.9)

Before trying to solve the Main Gap question it is natural to try to understand categoricity. The two (quite long) volumes, Shelah (2009a, 2009b), are a step in this direction. See in particular the introduction of Shelah (2009a); the works are in the context of AEC (abstract elementary classes).

While it is natural to repeat history and concentrate on the case of categoricity, can we say something *without* categoricity? In Shelah (undated a) it is suggested that we concentrate on cardinals $\lambda = \beth_\lambda$, i.e., fix points of the Beth sequence, which are of cofinality $\aleph_0$. The results are of the form: few models up to isomorphism implies every model in many such cardinals has arbitrarily large extensions.

Another direction is classifying theories T by the existence of non-isomorphic models of T of cardinality λ which are very equivalent by suitable long EF games (see Hyttinen and Tuuri 1991; Hyttinen *et al.* 1993; Hyttinen and Shelah 1994; Shelah 2008; and works in preparation). Hyttinen *et al.* (2013) have gone in a still different direction.

6 New logics

The model theory of first-order logic is relatively well understood, and inroads have been made into the model theory of various generalized quantifiers and infinitary logics. Still the following proposal seems timely:

> Find a new logic with good model theory (like compactness, completeness theorem, interpolation and those from 6.12) and strong expressive power preferably concerning other parts of mathematics (see Shelah (undated c), possibly specifically derive for them). (Shelah 2003a, Question 6.13)

On new model theoretically interesting logics, a step ahead seems the infinitary logic of Shelah (2012), which is between $\mathbb{L}_{<\kappa,\aleph_0} (= \cup_{\lambda<\kappa}\mathbb{L}_{\lambda,\aleph_0})$ and $\mathbb{L}_{<\kappa,<\kappa} (=\cup_{\lambda<\kappa} \mathbb{L}_{\lambda,\lambda})$ for $\kappa = \beth_\kappa$, and which seems susceptible to model theoretic treatment. For example, in these logics we cannot define well ordering (in a strong sense), we have addition of theories and product of two (all this like $\mathbb{L}_{<\kappa, \aleph_0}$ when $\kappa = \beth_\kappa$) but we also have the (long sought for) interpolation and, last but not least, a characterization, as Lindström had the celebrated characterization of first-order logic.

In other contexts we have explained that while advanced model theory, say classification theory, can be pursued for $\mathbb{L}_{\lambda, \aleph_0}$, this is not the case for $\mathbb{L}_{\aleph_1, \aleph_1}$ and stronger logics, because properties like categoricity are very sensitive to set theory. Note however that for example, monadic second-order logic on linear orders is not so affected. Still, it seemed to me that the arguments were quite conclusive. Having explained it (categoricity is sensitive to set theory) in a class, second thoughts have arisen. We have in mind the logic $\mathbb{L}_{\theta, \theta}$, with θ strongly compact. In Shelah (undated e) we have generalized the characterization of elementary equivalence in terms of isomorphic ultra-limits for the logic introduced in Shelah (2012), but the main point is generalizing the characterization of theories T such that for some theory T_1 in the logic $\mathbb{L}_{\theta, \theta}$ extending T the following holds: if M_1 is a model of T_1 then the τ_T-reduct of M_1 is saturated (reasonably defined). This seems to show that advanced model theory (e.g., classification theory) for the logic $\mathbb{L}_{\theta, \theta}$, θ a compact cardinal, is not empty. However, for categoricity the picture is opaque and quite different than the first-order case, for example, some categorical theories have models with long orders.

In another direction it seems interesting to find similar logics which would still have the upward Löwenheim–Skolem theorem. It seems best to find a maximal logic $\subseteq \mathbb{L}_{<\kappa, <\kappa}$ which is "nice" and satisfies the upward Löwenheim–Skolem theorem. It seems reasonable to base such a logic on Ehrenfeucht–Mostowski models. The proof in Shelah (undated a) seems relevant.

Also the question of the existence of a compact proper extension of first-order logic with interpolation is still unsolved.

As for compact logics related to algebra, the old papers Shelah (1994a, undated b) and the history given there, offer avenues forward but this has not materialized yet. A later paper with similar content is the unfinished Shelah (undated c), dealing with generalized quantifiers Q for which $\mathbb{L}(Q)$ is compact, expressing interesting properties. These works generally deal with (quantifying on automorphisms of) Boolean algebras, ordered fields and some triangle-free graphs. There is an undercurrent of connection to instability theory but no general theory has emerged so far. Also it is reasonable to assume that on finding such logics with quantifiers related to some specific mathematical field, proving compactness will have applications to such a field.

7 Abstract elementary classes

Is there any such thing as a pure semantic proof in model theory? Baldwin (2013) says about the author's Presentation Theorem that, passing through the syntax, we obtain a purely semantic theorem in the sense that we are able to deduce what he calls "purely semantical conclusions." We prefer to phrase it by saying that we can do model theory, i.e., investigate classes of models, without going through logic. We suppose this is the same (this means looking at AEC, see below).

The original aim in the investigation of abstract elementary classes (AEC), in Shelah (1975), was that we could strengthen the logic, but it was not clear how much. Originally it was $\mathbb{L}(Q)$, and then the infinitary version $\mathbb{L}_{\aleph_1, \aleph_0}(Q)$. But there was no maximal logic. So rather than trying to find the maximal logic, which seemed difficult, we wrote down the semantic assumptions that we needed. This became what is known as the study of AECs.

Type, in particular complete type over a model, is a central notion. Traditionally a type is defined as a set of formulas. However, passing from the syntactical approach to the semantical one in Shelah (1987b) (in 1986) the orbital type was introduced (Grossberg and Baldwin prefer to call this the Galois type).

Formulas and types are different notions; in AEC (orbital) types remain central but the parallel to formulas may, but so far has not, play a significant role.

In the 1960s it become clear that saturation is better than universal homogeneity because we can deal with one element at a time. Hence model homogeneity should be replaced by sequence homogeneity; this is still true for elementary classes. In Shelah (1987a) this problem was circumvented using enough stability (that is, definability of types), and so-called materializing a type (rather than realizing it). In Shelah (1987b) we succeeded in getting the best of the two worlds; well, when we have amalgamation, otherwise model homogeneous makes no sense. That is, we showed that model homogeneity is equivalent to saturation. Saturation means that we have to realize types of single elements, as in saturation, recalling that the type of an α-tuple is not determined by its restriction to finite sub-tuples. Moreover, these are not arbitrary types, just (complete) types over sub-models, in the right abstract sense.

8 Conclusion

One cannot emphasize enough the interest in trying to find a forcing-like method for Peano arithmetic with the potential of leading to proofs of independence from PA of even very "reasonable" arithmetical statements.

For set theory we have forcing, but it has led to the question how to decide what kind of new axioms we should add to ZFC in order to decide some or most of the statements left undecided by ZFC. In the face of the difficulty of this question, what we call "semi-axioms" suggest themselves. These are new axioms which have many consequences and lead to a deep and beautiful theory, but at the same time are reasonable and have "positive measure." The axiom $\mathbf{V} = \mathbf{L}$ would fulfil the first criterion but fails miserably on the second. On the other hand, the axiom "the continuum is weakly inaccessible" would be reasonable but has only few consequences.

The "California school," based on determinacy axioms, is to me unsatisfactory for several reasons that we have discussed above. The main point is that the fact that a statement, such as a determinacy axiom, solves everything really nicely is not a sufficient reason to call the statement a "true axiom." I also consider that the California school gives too much weight to success in descriptive set theory. While descriptive set theory is surely important, for me it is not "the center."

Other problems, arising perhaps from model theory or algebra, suggest different semi-axioms and universes that may be the nicest from the corresponding perspective. One such is WGCH (the weak generalized continuum hypothesis, i.e., $2^\lambda < 2^{\lambda^+}$).

Another potentially interesting direction is large continuum. I dream about a reasonable semi-axiom which implies that many of the established cardinal invariants of the continuum are (pairwise) distinct, and also determines their order. Understanding large continuum seems to require a theory of iterated forcing for a large continuum as versatile as the one we have for $2^{\aleph_0} = \aleph_2$ (or for $2^{\aleph_0} = \aleph_1$). In this connection I ask what would be the real significance of $2^{\aleph_0} = \aleph_{753}$, or of $2^{\aleph_0} = \aleph_{\omega^3+\omega+5}$, or is it so that all values of $2^{\aleph_0}$ which are $> \aleph_2$ are similar in some sense.

Despite some people's misgivings about the so-called unintuitive consequences, the Axiom of Choice, AC, is part of the standard axioms ZFC of set theory. We have proposed developing combinatorial set theory for universes with limited amount of choice, and here we comment more, particularly on further developments concerning this proposal. I have previously maintained that there is no serious model theory without choice, but my mind has changed. In particular we now have a proof, discussed above, of Morley's theorem in ZF. The Main Gap theorem seems much harder. In this direction we suggest studying "reasonable powers" as analogs of cardinals, and considering categoricity in reasonable powers.

In combinatorial set theory, considering pcf theory is interesting to me. In this respect we suggest focusing on the RGCH, the revised generalized continuum hypothesis which gives, for example, that for every $\lambda \geq \beth_\omega$ for every large enough $\kappa < \beth_\omega$ we have $\lambda^{[\kappa]} = \lambda$. This also applies to set theory with weak choice, and there have been some advances. Some generalize the pcf theorem by replacing "cofinal sequence of members of $\Pi\bar{a}$ modulo a filter" by "a sequence of subsets of the product which is increasing and cofinal."

Continuing on the topic of set theory without full choice, we suggest Ax_4, which says that "$[\lambda]^{\aleph_0}$ is well orderable for every cardinal λ." The thesis is that in this context "set theory is not so far from normal," the only reasons for non-well-orderability are the "$\mathcal{P}(\kappa)$ not well ordered" for regular κ or sometimes $\mathcal{P}(\mathcal{P}\kappa)$). A relevant universe is the following variant of Easton's model: for regular $\kappa > \aleph_0$ we add many subsets but not a well ordering. We suggest investigating $ZF + DC + Ax_4$ more systematically.

The so-called Main Gap theorem was for first-order theories, but we propose to prove a form of it for $\psi \in \mathbb{L}_{\lambda^+, \omega}$. The abstract elementary classes (AEC) provide a natural framework for the study of such sentences. The two volumes, Shelah (2009a, 2009b), make advances on categoricity in abstract elementary classes with the long term aim of proving a Main Gap theorem in this context.

Another approach to classifying theories, where recent progress has taken place, is by means of existence of winning strategies in transfinite EF games between large models of the theory.

It was proposed in Shelah (2003a) to find a new logic with good model theory (like compactness, completeness theorem, interpolation and those from 6.12) and strong expressive power preferably concerning other areas of mathematics. One interesting candidate has emerged recently, namely a logic between $\mathbb{L}_{<\kappa, \aleph_0} (= \cup_{\lambda<\kappa} \mathbb{L}_{\lambda, \aleph_0})$ and $\mathbb{L}_{<\kappa, <\kappa} (= \cup_{\lambda<\kappa} \mathbb{L}_{\lambda, \lambda})$ for $\kappa = \beth_\kappa$, with interpolation and a kind of Lindström characterization. This logic is under further investigation.

Despite earlier reservations on having any model theory for $\mathbb{L}_{\aleph_1, \aleph_1}$ and stronger logics, we have started to study $\mathbb{L}_{\theta, \theta}$, with θ strongly compact with some reasonable initial results. For example, we generalize the characterization of elementary classes such that for some elementary class in a larger vocabulary all the reducts to the original vocabulary are in the original class and are saturated. Also, for the new logic mentioned above, we get a characterization of the elementary equivalence in terms of isomorphic ultra-limits. This seems to show that advanced model theory (e.g., classification theory) for the logic $\mathbb{L}_{\theta,\theta}$, θ a compact cardinal, is possible.

The question of the existence of a compact proper extension of first-order logic with interpolation is still unsolved.

This chapter ends with a discussion on the question to what extent abstract elementary classes lead to purely semantic proofs in model theory. We claim that we can do model theory, i.e., investigate model classes, without going through logic.

Bibliography

Oliver Aberth, 1971. The failure in computable analysis of a classical existence theorem for differential equations, *Proc. Am. Math. Soc.* **30**, 151–156.

Uri Abraham and Menachem Magidor, 2010. Cardinal arithmetic. In Matthew Foreman and Akihiro Kanamori (editors), *Handbook of Set Theory*, volumes 1, 2, 3, pp. 1149–1227. Springer, Dordrecht.

John W. Addison, 1959. Some consequences of the axiom of constructibility, *Fundam. Mathematicae* **46**, 337–357.

Andrew Adler, 1969. Some recursively unsolvable problems in analysis, *Proc. Am. Math. Soc.* **22**, 523–526.

S. I. Adyan, 1957a. Unsolvability of some algorithmic problems in the theory of groups, *Trudy Moskov. Mat. Obšč.* **6**, 231–298 (in Russian).

   1957b. Finitely presented groups and algorithms, *Dokl. Akad. Nauk SSSR (N.S.)* **117**, 9–12 (in Russian).

David J. Anick, 1985. Diophantine equations, Hilbert series, and undecidable spaces, *Ann. Math.* **122**(1), 87–112.

Andrew Arana and Paolo Mancosu, 2012. On the relationship between plane and solid geometry, *Rev. Symbolic Logic* **5**(2), 294–353.

W. W. Armstrong, 1974. Dependency structures of database relationships, *Information Processing* **74**.

A. Arnauld and P. Nicole, 1964. *The Art of Thinking, or the Port Royal Logic*, translated by J. Dickoff and P. James. Bobbs-Merrill, Indianapolis, IN.

David Aspero, Paul Larson, and Justin Moore, 2013. Forcing axioms and the continuum hypothesis, *Acta Mathematica* **210**, 1–29.

Joan Bagaria, 2006. Axioms of generic absoluteness. In *Logic Colloquium 2002*, volume 27 of Lecture Notes in Logic, pp. 28–47. Association of Symbolic Logic, La Jolla, CA.

John Baldwin, 2009. *Categoricity*, volume 50 of University Lecture Series. American Mathematical Society, Providence, RI.

   2013. Formalization, primitive concepts, and purity, *Rev. Symbolic Logic* **6**, 87–128.

St. Banach and A. Tarski, 1924. Sur la décomposition des ensembles de points en parties respectivement congruentes, *Fundam. Mathematicae* **6**, 244–277.

Y. Bar-Hillel, editor, 1966. *Essays on the Foundations of Mathematics: Dedicated to A. A. Fraenkel on his seventieth anniversary*, 2nd edition. Magnum Press, Jerusalem.

Jon Barwise, editor, 1977. *Handbook of Mathematical Logic*, Studies in Logic and the Foundations of Mathematics, volume 90. North-Holland, Amsterdam.

P. Benacerraf, 1973. Mathematical truth, *J. Philos.* **70**, 661–680; also in Benacerraf and Putnam (1983), pp. 403–420.

P. Benacerraf and H. Putnam, editors, 1983. *Philosophy of Mathematics: Selected Readings*, 2nd edition. Cambridge University Press, Cambridge.

Robert Berger, 1966. The undecidability of the domino problem, *Mem. Am. Math. Soc.* **66**, 72.

P. Bernays, 1918. *Beiträge zur axiomatischen Behandlung des Logik-Kalküuls*, Habilitationsschrift, University of Göttingen.

E. Beth, 1959. *The Foundations of Mathematics; a Study in the Philosophy of Science.* North-Holland, Amsterdam.

Andreas Blass and Saharon Shelah, 1987. There may be simple $P\aleph_1$- and $P\aleph_2$-points and the Rudin–Keisler ordering may be downward directed, *Ann. Pure Appl. Logic* **33**, 213–243.

Vincent D. Blondel and John N. Tsitsiklis, 2000. A survey of computational complexity results in systems and control, *Automatica J. IFAC* **36**(9), 1249–1274.

Vincent D. Blondel, Olivier Bournez, Pascal Koiran, and John N. Tsitsiklis, 2001. The stability of saturated linear dynamical systems is undecidable, *J. Comput. System Sci.* **62**(3), 442–462.

L. M. Blumenthal, 1940. A paradox, a paradox, a most ingenious paradox, *Am. Math. Monthly* **47**, 346–353.

B. Bolzano, 1804. *Betrachtungen über einige Gegenstände der Elementargeometrie.* Karl Barth, Prague; translated in part by S. Russ in Ewald (1996), pp. 172–174.

1810. *Beyträge zu einer begründeteren Darstellung der Mathematik.* Caspar Widtmann, Prague; translated by S. Russ in Ewald (1996), pp. 174–224.

1817a. *Die drey Probleme der Rectification, der Complanation und der Cubirung, ohne Betrachtung des unendlich Kleinen, ohne die Annahme des Archimedes und ohne irgend eine nicht streng eweisliche Voraussetzung geloest; angleich als Probe einer ganzlichen Umstaltung der Raumwissenschaft allen Mathematikern zur Prüfung vorgelegt.* Gotthelf Kummer, Leipzig.

1817b. *Rein analytischer Beweis des Lehrsatzes, dass zwischen je zwei Werten, die ein entgegengesetztes Resultat gewähren, wenigstens eine reelle Wurzel der Gleichung liege.* Gottlieb Haase, Prague; translated by S. Russ in Ewald (1996), pp. 225–248.

1837. *Wissenschaftslehre. Versuch einer ausführlichen und grössenteils neuen Darstellung der Logik mit steter Rücksicht auf deren bisherigen Bearbeiter.* Seidel, Sulzbach; translated by B. Terrel in Bolzano (1973).

1973. *Theory of Science*, edited by J. Berg. D. Reidel, Boston, MA.

G. Boole, 1854. *An Investigation of the Laws of Thought.* Walton and Maberly, London.

George Boolos, 1971. The iterative conception of set, *J. Philos.* **68**(8), 215–231.

1998. Must we believe in set theory? In *Logic, Logic, and Logic*, pp. 120–132. Harvard University Press, Cambridge, MA.

William W. Boone, 1959. The word problem, *Ann. Math.* **70**(2), 207–265.

J. L. Borges, 1964. *Other Inquisitions 1937–1952*, translated by R. L. C. Simms. University of Texas Press, Austin, TX.

Robert Brandom, 2011. Platforms, patchworks, and parking garages: Wilson's account of conceptual fine-structure in *Wandering Significance*, *Philos. Phenomenol. Res.* **82**(1), 183–201.

L. E. J. Brouwer, 1913. Intuitionism and formalism, *Bull. Am. Math. Soc.* **20**, 81–96; reprinted in Benacerraf and Putnam (1983), pp. 77–89.

Dan Brumleve, Joel David Hamkins, and Philipp Schlicht, 2012. The mate-in-*n* problem of infinite chess is decidable (January 31, 2012). Preprint, arXiv:1201.5597v2.

Tyler Burge, 2005. Frege on sense and linguistic meaning. In *Truth, Thought, Reason*, pp. 242–269. Clarendon Press, Oxford.

John P. Burgess, 2008. *Mathematics, Models, and Modality*. Cambridge University Press.

J. Burgess and G. Rosen, 1997. Non-empirical physics. In *A Subject with No Object*, pp. 118–123. Oxford University Press, Oxford.

E. Carson, 2011. *Sensibility, Understanding and Number in Kant, Workshop on Knowledge, Representation and Proof in the Modern Era*, University of Notre Dame, November, 2011.

E. Carson and R. Huber, editors, 2006. *Intuition and the Axiomatic Method*, Western Ontario Series in Philosophy of Science. Springer, Dordrecht.

Alonzo Church, 1936a. An unsolvable problem of elementary number theory, *Am. J. Math.* **58**, 345–363.

    1936b. A note on the Entscheidungsproblem, *J. Symbolic Logic* **1**, 40–41.

Marcus Tullius Cicero, Treatise on Topics. In *The Orations of Marcus Tullius Cicero*, volume IV, pp. 1894–1903, translated by C. D. Yonge. Bell and Sons, London.

Paul Cohen, 1963. The independence of the continuum hypothesis, *Proc. Natl. Acad. Sci. USA* **50**, 1143–1148.

Paul J. Cohen, 1964. The independence of the continuum hypothesis. II, *Proc. Natl. Acad. Sci. USA* **51**, 105–110.

    1966. *Set Theory and the Continuum Hypothesis*. Benjamin, New York.

Donald J. Collins, 1972. Representation of Turing reducibility by word and conjugacy problems in finitely presented groups, *Acta Mathematica* **128** (1–2), 73–90.

A. Comte, 1892. *La Philosophie Positive: Les preliminaires généraux et la philosophie mathématique*, 5th edition. La Société Positiviste, Paris.

J. H. Conway, 1972. Unpredictable iterations, *Proceedings of the 1972 Number Theory Conference, University of Colorado, Boulder, CO, August 14–18, 1972*; reprinted in Lagarias (2010).

R. Courant and H. Robbins, 1981. *What is Mathematics?* Oxford University Press, Oxford.

George Csicsery, 2008. *Julia Robinson and Hilbert's tenth problem*. Documentary, Zala Films, http://zalafilms.com/films/juliarobinson.html.

Karel Culik II, 1996. An aperiodic set of 13 Wang tiles, *Discrete Math.* **160**(1–3), 245–251.

Martin Davis, 1953. Arithmetical problems and recursively enumerable predicates, *J. Symbolic Logic* **18**, 33–41.

1958. *Computability and Unsolvability*, McGraw-Hill Series in Information Processing and Computers. McGraw-Hill, New York.

editor, 1965. *The Undecidable: Basic papers on undecidable propositions, unsolvable problems and computable functions.* Raven Press, Hewlett, NY.

1977. Unsolvable problems. In Jon Barwise, editor, *Handbook of Mathematical Logic*, Studies in Logic and the Foundations of Mathematics, volume 90, pp. 567–594. North-Holland, Amsterdam.

editor, 2004. *The Undecidable: Basic papers on undecidable propositions, unsolvable problems and computable functions*, corrected reprint. Dover Publications, Mineola, NY.

Martin Davis, Hilary Putnam, and Julia Robinson, 1961. The decision problem for exponential diophantine equations, *Ann. Math.* **2** 74, 425–436.

Martin Davis, Yuri Matiyasevich, and Julia Robinson, 1976. Hilbert's tenth problem: Diophantine equations: positive aspects of a negative solution, *Mathematical Developments Arising from Hilbert Problems*, pp. 323–378 (loose erratum), Proc. Symp. Pure Math., De Kalb, IL, 1974, Vol. XXVIII. American Mathematical Society, Providence, RI.

J. W. Dawson, 1993. The compactness of first-order logic: from Gödel to Lindström, *History Philos. Logic* **14**(1), 15–37.

M. Dehn, 1911. Über unendliche diskontinuierliche Gruppen, *Math. Ann.* **71**(1), 116–144 (in German).

J. Denef, 1980. Diophantine sets over algebraic integer rings. II, *Trans. Am. Math. Soc.* **257**(1), 227–236.

J. Denef and L. Lipshitz, 1978. Diophantine sets over some rings of algebraic integers, *J. London Math. Soc.* 2 **18**(3), 385–391.

1984. Power series solutions of algebraic differential equations, *Math. Ann.* **267** (2), 213–238.

1989. Decision problems for differential equations, *J. Symbolic Logic* **54**(3), 941–950.

Jan Denef, Leonard Lipshitz, Thanases Pheidas, and Jan Van Geel, editors, 2000. *Hilbert's Tenth Problem: Relations with Arithmetic and Algebraic Geometry, Papers from the workshop held at Ghent University, Ghent, November 2–5, 1999*, Contemporary Mathematics, volume 270. American Mathematical Society, Providence, RI.

Michael Detlefsen, 1979. On interpreting Gödel's second theorem. *J. Philos. Logic*, **8**(3), 297–313.

2005. Formalism. In S. Shapiro (editor), *The Oxford Handbook of Philosophy of Mathematics and Logic*, Oxford University Press, Oxford.

Diogenes Laertius, 1925. *Lives of Eminent Philosophers*, translated by R. D. Hicks. Loeb Classical Library, Harvard University Press, Cambridge, MA.

John E. Doner, Andrzej Mostowski, and Alfred Tarski, 1978. The elementary theory of well-ordering—a metamathematical study. In *Logic Colloquium '77 (Proc. Conf., Wrocław, 1977)*, volume 96 of Studies in Logic and the Foundations of Mathematics, pp. 1–54. North-Holland, Amsterdam,.

M. Dummett, 1996. Frege and Kant on geometry. In M. Dummett (editor), *Frege and Other Philosophers*, pp. 126–158. Oxford University Press.

C. Duret, 1613. *Trésor de l'Histoire des Langues*. Cologne.

Kirsten Eisenträger, 2004. Hilbert's tenth problem for function fields of varieties over $\mathbb{C}$, *Int. Math. Res. Not.* **59**, 3191–3205.

Paul C. Eklof and Alan Mekler, 2002. *Almost Free Modules: Set Theoretic Methods*, volume 65 of North-Holland Mathematical Library, revised edition. North-Holland, Amsterdam.

Jörg Endrullis, Clemens Grabmayer, and Dimitri Hendriks, 2009. *Complexity of Fractran and Productivity, Automated Deduction*, CADE-22, Lecture Notes in Computer Science, volume 5663, pp. 371–387. Springer, Berlin.

W. Ewald, 1996. *From Kant to Hilbert: A Sourcebook in the Foundations of Mathematics*, volume I. Oxford University Press, New York.

S. Feferman, 1960–1961. Arithmetization of metamathematics in a general setting, *Fundam. Mathematicae* **49**, 35–92.

Solomon Feferman, M. Harvey, M. Friedman, Penelope Maddy, and John R. Steel, 2000. Does mathematics need new axioms? *Bull. Symbolic Logic* **6**(4), 401–446.

Solomon Feferman, Charles Parsons, and Stephen G. Simpson, editors, 2010. *Kurt Gödel: Essays for his Centennial*. Association for Symbolic Logic and Cambridge University Press.

Qi Feng, Menachem Magidor, and W. Hugh Woodin, 1992. Universally Baire sets of reals. In H. Judah, W. Just, and W. H. Woodin (editors), *Set Theory of the Continuum*, MSRI publications 26. Springer-Verlag, Berlin.

P. Finsler, 1926. Formale Beweise und die Entscheidbarkeit, *Math. Z.* **25**, 676–682.

J. Folina, 2010. Poincaré's philosophy of mathematics, *Internet Encyclopedia of Philosophy*.

M. Foucault, 1970. *The Order of Things: An Archaeology of the Human Sciences (a translation of Les Mots et les Choses)*. Random House, New York.

Curtis Franks, 2009. *The Autonomy of Mathematical Knowledge*. Cambridge University Press.

2010. Cut as consequence, *History Philos. Logic* **31**(4), 349–379.

G. Frege, 1980. *Philosophical and Mathematical Correspondence*, edited by G. Gabriel *et al.* Blackwell Publishers, Oxford.

1884. *Die Grundlagen der Arithmetik, 1884; references to The Foundations of Arithmetic*, translated by J. L. Austin. Northwestern University Press, Evanston, IL, 1980.

Chris Freiling, 1986. Axioms of symmetry: throwing darts at the real number line, *J. Symbolic Logic* **51**, 190–200.

M. Friedman, 1999. *Reconsidering Logical Positivism*. Cambridge University Press.

2001. *Dynamics of Reason.* Cambridge University Press.

2010. The a priori in physical theory (presented by Nick Huggett), Central Division Meeting, American Philosophical Association, Chicago, IL.

Pietro Galliani, 2012. Inclusion and exclusion dependencies in team semantics—on some logics of imperfect information, *Ann. Pure Appl. Logic* **163**(1), 68–84.

David Gamarnik, 2002. On deciding stability of constrained homogeneous random walks and queueing systems, *Math. Oper. Res.* **27**(2), 272–293.

2007. On the undecidability of computing stationary distributions and large deviation rates for constrained random walks, *Math. Oper. Res.* **32**(2), 257–265.

Robin Gandy, 1988. The confluence of ideas in 1936. In *The Universal Turing Machine: a half-century survey*, pp. 55–111. Oxford University Press, New York.

G. Gentzen, 1932. Über die Existenz unabhangiger Axiomensysteme zu unendlichen Satzsystemen, *Math. Ann.* **107**, 329–350; translated as On the existence of independent axiomsystems for infinite sentence systems, in Szabo (1969), pp. 29–52.

1934–1935. *Untersuchungen über das logische Schliessen*, Doctoral thesis, University of Gottingen; translated as Investigations into logical deduction, in Szabo (1969), pp. 68–131.

R. George, 1971. Editor's introduction. In B. Bolzano, *Theory of Science.* University of California Press, Los Angeles, CA.

J. Gergonne, 1826–1827. Géométrie de situation, *Ann. Math. Pures Appl.* **17**, 214–252.

Victoria Gitman and Joel Hamkins, 2010. A natural model of the multiverse axioms, *Notre Dame J. Formal Logic* **51**(4), 475–484.

Kurt Gödel, 1929. *Über die Vollständigkeit des Logikkalküls*, Doctoral thesis, University of Vienna; translated by S. Bauer-Mengelberg and Jean van Heijenoort, On the completeness of the calculus of logic, reprinted in Gödel (1986), pp. 60–101.

1930a. Über die Vollständigkeit des Logikkalküls, *Naturwissenschaften* **18**, 1068; reprinted with English translation in Gödel (1986).

1930b. Einige metamathematische Resultate über Entscheidungsdefinitheit und Widerspruchsfreiheit, *Anz. Akad. Wiss. Wien* **67**, 214–215; reprinted with English translation in Gödel (1986).

1930c. Die Vollständigkeit der Axiome des logischen Funktionenkalküls, *Monatsh. Math. Phys.* **37**(1), 349–360; reprinted with English translation in Gödel (1986), pp. 103–123.

1931a. Diskussion zur Grundlegung der Mathematik, *Erkenntnis* **2**, 147–151; reprinted with English translation in Gödel (1986).

1931b. Über formal unentscheidbare Sätze der *Principia mathematica* und verwandter Systeme I, *Monatsh. Math. Phys.* **38**, 173–198; reprinted with English translation in Gödel (1986).

1940. *The Consistency of the Continuum Hypothesis*, Annals of Mathematics Studies, number 3. Princeton University Press, Princeton, NJ.

1944. Russell's mathematical logic. In P. A. Schilpp (editor), *The Philosophy of Bertrand Russell*, pp. 123–153. Northwestern University, Evanstown, IL. Reprinted in Benacerraf and Putnam (1983), pp. 447–469.

1946. Remarks before the Princeton Bicentennial Conference of problems in mathematics, 1946, lecture first published in Davis (1965), and in Gödel (1990).

1947. What is Cantor's continuum problem? *Am. Math. Monthly* **54**, 515–525; reprinted in Benacerraf and Putnam (1983), pp. 470–485.

1949a. An example of a new type of cosmological solutions of Einstein's field equations of gravitation, *Rev. Mod. Phys.* **21**, 447–450.

1949b. A remark about the relationship between relativity theory and idealistic philosophy. In P. Schilpp (editor), *Albert Einstein: Philosopher Scientist*, pp. 555–562. Library of Living Philosophers, Evanston, IL.

1961. The modern development of the foundations of mathematics in the light of philosophy, lecture published in Gödel (1995), pp. 374–387.

1986. *Collected Works*, volume I, *Publications 1929–1936*, edited by S. Feferman, J. W. Dawson, Jr., S. C. Kleene, G. H. Moore, R. N. Solovay, and J. van Heijenoort. Oxford University Press, Oxford.

1990. *Collected Works*, volume II, *Publications 1938–1974*, edited by S. Feferman, J. W. Dawson, Jr., S. C. Kleene, G. H. Moore, R. N. Solovay, and J. van Heijenoort. Oxford University Press, Oxford.

1995. *Collected Works*, volume III, *Unpublished Essays and Lectures*, edited by S. Feferman, J. W. Dawson, Jr., W. Goldfarb, C. Parsons, and R. N. Solovay. Oxford University Press, Oxford.

2003a. *Collected Works*, volume IV, *Correspondence A–G*, edited by S. Feferman, J. W. Dawson, Jr., W. Goldfarb, C. Parsons, and W. Sieg. Oxford University Press, Oxford.

2003b. *Collected Works*, volume V, *Correspondence H–Z*, edited by S. Feferman, J. W. Dawson, Jr., W. Goldfarb, C. Parsons, and W. Sieg. Oxford University Press, Oxford.

A. W. Goodman, 1959. On sets of acquaintances and strangers at any party, *Am. Math. Monthly* **66**, 778–783.

R. L. Goodstein, 1953. *On the Metamathematics of Algebra* by A. Robinson, *Mathematical Gazette* **37**(321), 224–226.

Erich Grädel and Jouko Väänänen, 2013. Dependence and independence, *Studia Logica* **101**(2), 233–236.

Jacques Hadamard, 1945. *The Psychology of Invention in the Mathematical Field.* Dover, New York.

Hans Hahn, 1956. The crisis in intuition. In James R. Newman (editor), *The World of Mathematics*, volume III, *1956–1976*. Simon & Schuster, New York.

Wolfgang Haken, 1961. Theorie der Normalflächen, *Acta Mathematica* **105**, 245–375 (in German).

Vesa Halava and Tero Harju, 2001. Mortality in matrix semigroups, *Am. Math. Monthly* **108**(7), 649–653.

Vesa Halava, Tero Harju, Mika Hirvensalo, and Juhani Karhumäki, 2005. *Skolem's problem – on the border between decidability and undecidability*, TUCS Technical Reports, no. 683, Turku Centre for Computer Science, April 2005. Available at: http://tucs.fi/publications/view/?pub_id=tHaHaHiKao5a.

Vesa Halava, Tero Harju, and Mika Hirvensalo, 2007. Undecidability bounds for integer matrices using Claus instances, *Int. J. Found. Comput. Sci.* **18**(5), 931–948.

M. Hallett, 2006. Gödel, realism and mathematical "intuition". In E. Carson and R. Huber (editors), *Intuition and the Axiomatic Method*, Western Ontario Series in Philosophy of Science, pp. 113–131. Springer, Dordrecht.

Joel Hamkins, 2011. The set-theoretic multiverse: a natural context for set theory (mathematical logic and its applications), *Ann. Jpn. Assoc. Philos. Sci.* **19**, 37–55.

Joel Hamkins, Jonas Reitz, and W. Hugh Woodin, 2007. The ground axiom is consistent with *V*=HOD, *Proc. Am. Math. Soc.* I.

Michael Harris, to appear. *Not Merely Good, True, and Beautiful*. Princeton University Press, Princeton, NJ.

Hamed Hatami and Serguei Norine, 2011. Undecidability of linear inequalities in graph homomorphism densities, *J. Am. Math. Soc.* **24**(2), 547–565.

Kai Hauser, 2006. Gödel's program revisited, part I: the turn to phenomenology, *Bull. Symbolic Logic* **12**, 529–590.

Robert Aubrey Hearn, 2006. *Games, puzzles, and computation*, Ph.D. thesis, Department of Electrical Engineering and Computer Science, Massachusetts Institute of Technology. Available at www.swiss.ai.mit.edu/~bob/hearn-thesis-final.pdf.

Robert A. Hearn and Erik D. Demaine, 2009. *Games, Puzzles, and Computation*. A K Peters, Wellesley, MA.

Geoffrey Hemion, 1979. On the classification of homeomorphisms of 2-manifolds and the classification of 3-manifolds, *Acta Mathematica* **142**(1–2), 123–155.

L. Henkin, 1961. Some remarks on infinitely long formulas. In *Infinitistic Methods (Proc. Symp. Foundations of Mathematics, Warsaw, 1959)*, pp. 167–183. Pergamon, Oxford.

P. Hertz, 1929. Über Axiomensysteme für beliebige Satzsysteme, *Math. Ann.* **101**, 457–514.

A. Heyting, 1963. *Axiomatic Projective Geometry*. Noordhoff, Groningen.

1966. Axiomatic method and intuitionism. In Y. Bar-Hillel (editor), *Essays on the Foundations of Mathematics: Dedicated to A. A. Fraenkel on his seventieth anniversary*, 2nd edition, pp. 237–247. Magnum Press, Jerusalem.

G. Higman, 1961. Subgroups of finitely presented groups, *Proc. R. Soc. London, Ser. A* **262**, 455–475.

D. Hilbert, 1900a. Mathematische Probleme, *Göttinger Nachr.* 253–297; translated by M. W. Newson in *Bull. Am. Math. Soc.*, **8**, 1902, 437–479.

1900b. Über den Zahlbegriff, *Jahresber. Deutsch. Math. Verein.* **8**, 180–183; reprinted in Hilbert (1913).

1913. *Grundlagen der Geometrie*, 4th edition. Teubner, Leipzig.

1930a. Königsberg radio address. Available at: http://math.sfsu.edu/smith/Documents/HilbertRadio/HilbertRadio.mp3.

1930b. Probleme der Grundlegung der Mathematik, *Math. Ann.* **102**, 1–9.

D. Hilbert and W. Ackermann, 1928. *Grundzüge der Theoretischen Logik*. Springer, Berlin; translation in Luce (1950).

M. Holz, K. Steffens, and E. Weitz, 1999. *Introduction to Cardinal Arithmetic*, Birkhäuser Advanced Texts. Birkhäuser, Basel.

David Hume, 1751. *Enquiry Concerning the Principles of Morals*, §III, part II, ¶ 10, edited by J. Schneewind. Hackett, Indianapolis, IN, 1983.

E. Huntington, 1911. The fundamental propositions of algebra. In J. W. Young (editor), *Lectures on Fundamental Concepts of Algebra and Geometry*, Chapter IV, pp. 151–210. Macmillan, New York.

Tapani Hyttinen and Saharon Shelah, 1994. Constructing strongly equivalent nonisomorphic models for unsuperstable theories, part A, *J. Symbolic Logic* **59**, 984–996.

Tapani Hyttinen and Heikki Tuuri, 1991. Constructing strongly equivalent nonisomorphic models for unstable theories, *Ann. Pure Appl. Logic* **52**, 203–248.

Tapani Hyttinen, Saharon Shelah, and Heikki Tuuri, 1993. Remarks on strong nonstructure theorems, *Notre Dame J. Formal Logic* **34**, 157–168.

Tapani Hyttinen, Kaisa Kangas, and Jouko Väänänen, 2013. On second order characterizability, *Logic J. IGPL* **21**, 767–787.

Gérard Jacob, 1977–1978. Un algorithme calculant le cardinal, fini ou infini, des demi-groupes de matrices, *Theor. Comput. Sci.* **5**(2), 183–204 (in French, with English summary).

1978. La finitude des représentations linéaires des semi-groupes est décidable, *J. Algebra* **52**(2), 437–459 (in French, with English summary).

S. Jaśkowski, 1954. Example of a class of systems of ordinary differential equations having no decision method for existence problems, *Bull. Acad. Polon. Sci. Cl. III* **2**, 155–157.

J. P. Jones, 1982. Some undecidable determined games, *Int. J. Game Theory* **11**(2), 63–70.

Lázló Kalmár, 1928–1929. Zur Theorie der abstrakten Spiele, *Acta Sci. Math. Szeged* **4**, 65–85; English translation, *The Foundations of Game Theory*, volume I, 1997, pp. 247–262.

I. Kant, 1783. *Prolegomena to any Future Metaphysics*, translated by Paul Carus and James Ellington. Hackett, Indianapolis, IN, 1977.

1787. *Critique of Pure Reason*, translated by Norman Kemp Smith. St Martin's Press, New York, 1965.

1992. The Jäsche logic. In J. M. Young, editor, *Lectures on Logic*. Cambridge University Press.

Jarkko Kari, 1996. A small aperiodic set of Wang tiles, *Discrete Math.* **160**(1–3), 259–264.

Jakob Kellner and Saharon Shelah, 2009. Decisive creatures and large continuum, *J. Symbolic Logic* **74**, 73–104.

Juliette Kennedy, 2011. Gödel's thesis: an appreciation. In Mathias Baaz (editor), *Kurt Gödel and the Foundations of Mathematics: Horizons of Truth*. Cambridge University Press, Cambridge.

2013. On formalism freeness: implementing Gödel's 1946 Princeton bicentennial lecture, *Bull. Symbolic Logic* **19**, 351–393.

K. H. Kim and F. W. Roush, 1992. Diophantine undecidability of $\mathbb{C}(t_1, t_2)$, *J. Algebra* **150**(1), 35–44.

David A. Klarner, Jean-Camille Birget, and Wade Satterfield, 1991. On the undecidability of the freeness of integer matrix semigroups, *Int. J. Algebra Comput.* **1**(2), 223–226.

S. C. Kleene, 1936. General recursive functions of natural numbers, *Math. Ann.* **112**(1), 727–742.

M. Kline, 1972. *Mathematical Thought from Ancient to Modern Times*. Oxford University Press, New York.

G. Kneebone, 1963. *Mathematical Logic and the Foundations of Mathematics*. D. van Nostrand, London.

Peter Koellner, 2006. On the question of absolute undecidability, *Philos. Math.* **3** (14), 153–188; reprinted with revisions in Feferman *et al.* (2010).

2009. On reflection principles, *Ann. Pure Appl. Logic* **157**(2–3), 206–219.

Peter Koellner and W. Hugh Woodin, 2010. Large cardinals from determinacy. In M. Foreman and A. Kanamori (editors), *Handbook of Set Theory*, volume 3, pp. 1951–2119. Springer, Berlin.

Jochen Koenigsmann, 2010. Defining $\mathbb{Z}$ in $\mathbb{Q}$. Preprint, arXiv:1011.3424.

János Kollár, 2008. Diophantine subsets of function fields of curves, *Algebra Number Theory* **2**(3), 299–311.

Dénes König, 1927. Über eine Schlussweise aus dem Endlichen ins Unendliche, *Acta Sci. Math. Szeged* **3**, 121–130 (in German).

Juha Kontinen and Jouko Väänänen, 2013. Axiomatizing first order consequences In dependence logic. *Ann Pure Appl. Logic* **164**, 1101–1117.

V. Krebs and W. Day, editors, 2010. *Seeing Wittgenstein Anew: New Essays on Aspect Seeing*. Cambridge University Press, Cambridge.

Georg Kreisel, 1967. Informal rigour and completeness proofs. In *Proceedings of the International Colloquium in the Philosophy of Science, London, 1965*, edited by Imre Lakatos, volume 1, pp 138–157. North-Holland, Amsterdam.

1983. Hilbert's programme. In Paul Benacerraf and Hilary Putnam, editors, *Philosophy of Mathematics: Selected Readings*, 2nd edition, pp. 207–238. Cambridge University Press.

1987. Gödel's excursions into intuitionistic logic. In *Gödel Remembered (Salzburg, 1983)*, History of Logic, IV, pp. 65–186. Bibliopolis, Naples.

S. Kripke, 1980. *Naming and Necessity*. Harvard University Press, Cambridge, MA.

Stuart A. Kurtz and Janos Simon, 2007. The Undecidability of the Generalized Collatz Problem, *Theory and Applications of Models of Computation*, Lecture Notes in Computer Science, volume 4484, pp. 542–553. Springer, Berlin.

M. Laczkovich, 2003. The removal of π from some undecidable problems involving elementary functions, *Proc. Am. Math. Soc.* **131**(7), 2235–2240.

Jeffrey C. Lagarias, 1985. The 3x+1 problem and its generalizations, *Am. Math. Monthly* **92**(1), 3–23.

editor, 2010. *The Ultimate Challenge: the 3x + 1 Problem.* American Mathematical Society, Providence, RI.

J. L. Lagrange, 1788. *Mécanique Analytique.* Desaint, Paris.

1797. *Théorie des Fonctions Analytiques.* Imprimerie de la Republique, Paris.

Serge Lang, 1966. *Introduction to Transcendental Numbers.* Addison-Wesley, Reading, MA.

C. H. Langford, 1927. On a type of completeness characterizing the general laws for separation of point-pairs, *Trans. Am. Math. Soc.* **29**, 96–110.

P. S. Laplace, 1813. *Exposition du Système du Monde.* Mme ve Courcier, Paris.

Paul Larson, 2004. *The Stationary Tower: Notes on a Course by W. Hugh Woodin.* American Mathematical Society, Providence, RI.

Paul Larson and Saharon Shelah, 2009. Splitting stationary sets from weak forms of choice, *Math. Logic Q.* **55**, 299–306.

Richard Laver, 2007. Certain very large cardinals are not created in small forcing extensions, *Ann. Pure Appl. Logic* **149**(1–3), 1–6.

F. W. Lawvere, 1964. An elementary theory of the category of sets, *Proc. Natl. Acad. Sci. USA* **52**, 1506–1511.

A. Lévy, 1960–1961. Principles of reflection in axiomatic set theory, *Fundam. Mathematicae* **49**, 1–10.

Azriel Levy and Robert Solovay, 1967. Measurable cardinals and the continuum hypothesis, *Isr. J. Math.* **5**, 233–248.

Per Lindström, 2000. Quasi-realism in mathematics, *Monist* **83**(1), 122–149.

Joseph Liouville, 1833. Premier et second mémoire sur la détermination des intégrales dont la valeur est algébrique, *J. l'École Polytech.* **14**, 124–148 and 149–193.

1835. Mémoire sur l'intégration d'une classe de fonctions transcendentes, *J. reine angew. Math.* **13**(2), 93–118.

L. Löwenheim, 1915. Über Möglichkeiten im Relativkalkül, *Math. Ann.* **76**, 447–470.

Robert E. Luce, editor, 1950. *Principles of Mathematical Logic.* Steinhardt, Chelsea, NY.

Angus Macintyre and A. J. Wilkie, 1996. *On the Decidability of the Real Exponential Field,* pp. 441–467. Kreiseliana, A. K. Peters, Wellesley, MA.

Penelope Maddy, 1990. *Realism in Mathematics.* Oxford University Press, Oxford.

1997. *Naturalism in Mathematics.* Clarendon Press, Oxford.

1998a. V = L and MAXIMIZE. In *Logic Colloquium 1995 (Haifa)*, volume 11 of Lecture Notes in Logic, pp. 134–152. Springer, Berlin.

1998b. Believing the axioms, *J. Symbolic Logic* **53**, 481–511 and 736–764.

2009. *Second Philosophy.* Oxford University Press, Oxford.

2011. *Defending the Axioms.* Oxford University Press, Oxford.

Arnaldo Mandel and Imre Simon, 1977–1978. On finite semigroups of matrices, *Theoret. Comput. Sci.* **5**(2), 101–111.

Benoît Mandelbrot, 1977. *The Fractal Geometry of Nature.* W. H. Freeman, New York.

Maurice Margenstern, 2008. The domino problem of the hyperbolic plane is undecidable, *Theoret. Comput. Sci.* **407**(1–3), 29–84.

A. Markov, 1947. The impossibility of certain algorithms in the theory of associative systems, II, *Dokl. Akad. Nauk SSSR (N.S.)* **58**, 353–356 (in Russian).

1951. The impossibility of certain algorithms in the theory of associative systems, *Dokl. Akad. Nauk SSSR (N.S.)* **77**, 19–20 (in Russian).

1958. The insolubility of the problem of homeomorphy, *Dokl. Akad. Nauk SSSR* **121**, 218–220 (in Russian).

Donald Martin, 1976. Hilbert's first problem: the continuum hypothesis. In F. Browder (editor), *Mathematical Developments Arising from Hilbert Problems*, Proceedings of Symposia in Pure Mathematics, volume 28, pp. 81–92. American Mathematical Society, Providence, RI.

1998. Mathematical evidence. In H. G. Dales and G. Oliveri (editors), *Truth in Mathematics*, pp. 215–231. Clarendon Press, Oxford.

2005. Gödel's conceptual realism, *Bull. Symbolic Logic* **2**, 207–224.

Donald Martin and Robert Solovay, 1970. Internal Cohen extensions, *Ann. Math. Logic* **2**, 143–178.

Yuri V. Matiyasevich, 1970. The Diophantineness of enumerable sets, *Dokl. Akad. Nauk SSSR* **191**, 279–282 (in Russian).

1993. *Hilbert's Tenth Problem*, Foundations of Computing Series. MIT Press, Cambridge, MA.

Hideyuki Matsumura, 1963. On algebraic groups of birational transformations, *Atti Accad. Naz. Lincei Rend. Cl. Sci. Fis. Mat. Natur.* **34**(8), 151–155.

B. Mazur, 1994. Questions of decidability and undecidability in number theory, *J. Symbolic Logic* **59**(2), 353–371.

Barry Mazur and Karl Rubin, 2010. Ranks of twists of elliptic curves and Hilbert's tenth problem, *Invent. Math.* **181**, 541–575.

Kenneth McAloon, 1970–1971. Consistency results about ordinal definability, *Ann. Math. Logic* **2**(4), 449–467.

G. S. Mendick and J. K. Truss, 2003. A notion of rank in set theory without choice, *Arch. Math. Logic* **42**(2), 165–178.

Charles F. Miller III, 1992. *Decision Problems for Groups – Survey and Reflections, Algorithms and Classification in Combinatorial Group Theory (Berkeley, CA, 1989)*, Math. Sci. Res. Inst. Publ., volume 23, pp. 1–59. Springer, New York.

D. Mirimanoff, 1917. Les antinomies de Russell et de Burali-Forti et le problème fondamental de la théorie des ensembles, *Enseign. Math.* **19**, 37–52.

William Mitchell, 1979. Hypermeasurable cardinals. In Maurice Boffa, Dirk van Dalen, and Kenneth McAloon (editors), *Logic Colloquium '78, Proceedings of the Colloquium held in Mons, August 1978*, pp. 303–316.

Cristopher Moore, 1990. Unpredictability and undecidability in dynamical systems, *Phys. Rev. Lett.* **64**(20), 2354–2357.

Gregory H. Moore, 1988. The emergence of first-order logic. In William Aspray and Philip Kitcher (editors), *History and Philosophy of Modern Mathematics*, Minnesota Studies in the Philosophy of Science, volume 11, pp. 95–135. University of Minnesota Press, Minneapolis, MN.

Justin Tatch Moore, 2013. Forcing axioms and the continuum hypothesis, part II: transcending ω_1-sequences of reals, *Acta Mathematica* **210**, 173–183.

Laurent Moret-Bailly, 2005. Elliptic curves and Hilbert's tenth problem for algebraic function fields over real and *p*-adic fields, *J. reine angew. Math.* **587**, 77–143.

Yiannis Moschovakis, 1981. Ordinal games. In A. S. Kechris, D. A. Martin, and Y. N. Moscovakis (editors), *Cabal Seminar 77–79*, Lecture Notes in Mathematics, volume 839, pp. 169–201. Springer-Verlag, Berlin.

2009. *Descriptive Set Theory*, 2nd edition. American Mathematical Society, Providence, RI.

Andrzej Mostowski, 1964. Widerspruchsfreiheit und Unabhängigkeit der Kontinuumhypothese, *Elem. Math.* **19**, 121–125.

John Myhill and Dana Scott, 1971. Ordinal definability. In D. Scott (editor), *Axiomatic Set Theory*, Proceedings of Symposia in Pure Mathematics, volume 13, part 1, pp. 271–278. American Mathematical Society, Providence, RI.

Alexander Nabutovsky and Shmuel Weinberger, 1996. Algorithmic unsolvability of the triviality problem for multidimensional knots, *Comment. Math. Helv.* **71**(3), 426–434.

Itay Neeman, 2007. Games of length ω_1, *J. Math. Logic*, **7**, 831–824.

P. S. Novikov, 1954. Unsolvability of the conjugacy problem in the theory of groups, *Izv. Akad. Nauk SSSR. Ser. Mat.* **18**, 485–524 (in Russian).

1955. Ob algoritmičeskoĭ nerazrešimosti problemy tožestva slov v teorii grupp, *Tr. Mat. Inst. im. Steklov., Izdat. Akad. Nauk SSSR*, **44** (in Russian); English translation, On the algorithmic insolvability of the word problem in group theory, *Am. Math. Soc. Transl.* **9**(2), 1958.

R. J. Nunke, 1977. Whitehead's problem. In *Abelian Group Theory (Proc. Second New Mexico State Univ. Conf.)*, volume 616 of Lecture Notes in Mathematics, pp. 240–250. Springer, Berlin.

John C. Oxtoby, 1971. *Measure and Category*. Springer, Berlin.

Pappus, 1941. The treasury of analysis. In I. Thomas (editor), *Selections Illustrating the History of Greek Mathematics*, volume II, *From Aristarchus to Pappus*, pp. 596–599. Loeb Classical Library, Harvard University Press, Cambridge, MA.

Jennifer Park, 2012. A universal first order formula defining the ring of integers in a number field, February 28. Preprint, arXiv:1202.6371v1.

Charles Parsons, 1977. What is the iterative conception of set? In R. E. Butts and Jaako Hintikka (editors), *Logic, Foundations of Mathematics and Computability Theory (Proc. Fifth Int. Congress on Logic, Methodology and Philosophy of Science, University of Western Ontario, 1975), Part I*, University of Western Ontario Series on the Philosophy of Science, volume 9, pp. 335–367. Reidel, Dordrecht.

1980. Mathematical intuition, *Proc. Aristotelian Soc.*, pp. 142–168.

1983. *Mathematics in Philosophy*. Cornell University Press, Ithaca, NY.

1992. The impredicativity of induction. In Michael Detlefsen (editor), *Proof, Logic, and Formalization*. Routledge, London.

1995a. Platonism and mathematical intuition in Kurt Gödel's thought, *Bull. Symbolic Logic* **1**, 44–74.

1995b. Quine and Gödel on analyticity. In Paolo Leonardi and Marco Santambrogio (editors), *On Quine: New Essays*, pp. 297–313. Cambridge University Press.

1998. Hao Wang as philosopher and interpreter of Gödel, *Philos. Math. 3* **6**(3–24), 9–11; reprinted in Parsons (forthcoming).

2008. *Mathematical Thought and its Objects*. Cambridge University Press.

2010. Gödel and philosophical idealism, *Philos. Math.* **18**, 166–192.

2014. *Philosophy of Mathematics in the Twentieth Century*. Harvard University Press, Cambridge, MA.

Michael S. Paterson, 1970. Unsolvability in 3 × 3 matrices, *Stud. Appl. Math.* **49**, 105–107.

C. S. Peirce, 1898. The logic of mathematics in relation to education, *Educ. Rev.* **15**, 209–216.

Thanases Pheidas, 1988. Hilbert's tenth problem for a class of rings of algebraic integers, *Proc. Am. Math. Soc.* **104**(2), 611–620.

S. Pinker, 1994. *The Language Instinct*. William Morrow, New York.

H. Poincaré, 1902. *La Science et l'Hypothèse*. Flammarion, Paris; translated by G. B. Halsted, *Science and Hypothesis*, The Science Press, New York, 1905.

1913. *Dernières Pensées*. Flammarion, Paris; translated by J. W. Bolduc, *Mathematics and Science: Last Essays*, Dover, New York, 1963.

George Polya, 1954. *Mathematics and Plausible Reasoning*, volume I, *Induction and Analogy in Mathematics*, volume II, *Patterns of Plausible Inference*. Princeton University Press, Princeton, NJ.

Bjorn Poonen, 2002. Using elliptic curves of rank one towards the undecidability of Hilbert's tenth problem over rings of algebraic integers. In *Algorithmic Number Theory*, pp. 33–42. Springer, Berlin.

2008. Undecidability in number theory, *Notices Am. Math. Soc.* **55**(3), 344–350.

2009. Characterizing integers among rational numbers with a universal-existential formula, *Am. J. Math.* **131**(3), 675–682.

2011. Automorphisms mapping a point into a subvariety, *J. Algebraic Geom.* **20**(4), 785–794.

Emil. L. Post, 1921. Introduction to a general theory of elementary propositions, *Am J. Math.* **43**, 163–185; reprinted in Van Heijenoort (1967a), pp. 265–282.

1944. Recursively enumerable sets of positive integers and their decision problems, *Bull. Am. Math. Soc.* **50**, 284–316.

1946. A variant of a recursively unsolvable problem, *Bull. Am. Math. Soc.* **52**, 264–268.

1947. Recursive unsolvability of a problem of Thue, *J. Symbolic Logic* **12**, 1–11.

Marian Boykan Pour-El and Ian Richards, 1979. A computable ordinary differential equation which possesses no computable solution, *Ann. Math. Logic* **17**(1–2), 61–90.

1983. Noncomputability in analysis and physics: a complete determination of the class of non-computable linear operators, *Adv. Math.* **48**(1), 44–74.

Mojzesz Presburger, 1929. Über die Vollständigkeit eines gewissen Systems der Arithmetik ganzer Zahlen, in welchem die Addition als einzige Operation hervortritt, *Comptes Rendus du I congrès de Mathématiciens des Pays Slaves, Warsaw*, pp. 92–101 (in German); translated by Dale Jacquette, On the completeness of a certain system of arithmetic of whole numbers in which addition occurs as the only operation, *Hist. Philos. Logic*, **12**, 1991, 225–233.

Pavel Pudlák, 1996. On the lengths of proofs of consistency: a survey of results. In *Collegium Logicum*, volume 2 of Coll. Logicum Ann. Kurt-Gödel-Soc., pp. 65–86. Springer, Vienna.

Michael O. Rabin, 1957. Effective computability of winning strategies. *Contributions to the Theory of Games*, volume 3, Annals of Mathematics Studies, no. 39, pp. 147–157. Princeton University Press, Princeton, NJ.

1958. Recursive unsolvability of group theoretic problems, *Ann. Math. 2* **67**, 172–194.

K. Reidemeister, 1930. *Vorlesungen über Grundlagen der Geometrie*. Springer, Berlin.

W. N. Reinhardt, 1974. Remarks on reflection principles, large cardinals, and elementary embeddings. In *Axiomatic Set Theory*, part 2, Proceedings of Symposia in Pure Mathematics, volume 13, part 2, pp. 189–205. American Mathematical Society, Providence, RI.

Jonas Reitz, 2007. The ground axiom, *J. Symbolic Logic*, **72**(4), 1299–1317.

Glenn C. Rhoads, 2005. Planar tilings by polyominoes, polyhexes, and polyiamonds, *J. Comput. Appl. Math.* **174**(2), 329–353.

Daniel Richardson, 1968. Some undecidable problems involving elementary functions of a real variable, *J. Symbolic Logic* **33**, 514–520.

Assaf Rinot, 2007. Aspects of singular cofinality, *Contrib. Discrete Math.* **2**(2), 185–204.

Robert H. Risch, 1970. The solution of the problem of integration in finite terms, *Bull. Am. Math. Soc.* **76**, 605–608.

A. Robinson, 1951. *On the Metamathematics of Algebra*. North-Holland, Amsterdam.

Julia Robinson, 1949. Definability and decision problems in arithmetic, *J. Symbolic Logic* **14**, 98–114.

Raphael M. Robinson, 1971. Undecidability and nonperiodicity for tilings of the plane, *Invent. Math.* **12**, 177–209.

1978. Undecidable tiling problems in the hyperbolic plane, *Invent. Math.* **44**(3), 259–264.

W. Rohloff, 2010. Kantian intuition and the applicability of Geometry, *Midwest Philosophy of Mathematics Workshop, University of Notre Dame, October, 2010*.

Maxwell Rosenlicht, 1972. Integration in finite terms, *Am. Math. Monthly* **79**, 963–972.

Barkley Rosser, 1936. Extensions of some theorems of Gödel and Church, *J. Symbolic Logic* **1**, 87–91.

Lee A. Rubel, 1981. A universal differential equation, *Bull. Am. Math. Soc. (N.S.)* **4**(3), 345–349.

1983. Some research problems about algebraic differential equations, *Trans. Am. Math. Soc.* **280**(1), 43–52.

1992. Some research problems about algebraic differential equations. II, *Illinois J. Math.* **36**(4), 659–680.

M. Ruse, 1998. *Taking Darwin Seriously.* Prometheus Books, Amherst, NY.

B. Russell, 1907. The Regressive Method of Discovering the Premises of Mathematics, lecture to the Cambridge Mathematical Club, March 9, 1907. Printed in D. Lackey (editor), *Essays in Analysis: Bertrand Russell.* George Braziller, New York, 1973.

Grigor Sargsyan, 2009. *A tale of hybrid mice*, Ph.D. Thesis, University of California, Berkeley, CA.

P. Schilpp, 1951. *The Philosophy of Bertrand Russell*, Library of Living Philosophers, volume 5, 3rd edition. Tudor Press, New York.

Joseph R. Schoenfield, 1959. On the independence of the axiom of constructibility, *Am. J. Math.* **81**(3), 537–540.

Dana Scott, 1961. Measurable cardinals and constructible sets, *Bull. Acad. Polon. Sci. Ser. Sci. Math. Astronom. Phys.* **9**, 521–524.

Paul Seidel, 2008. A biased view of symplectic cohomology, *Current Developments in Mathematics, 2006*, pp. 211–253. International Press, Somerville, MA.

Saharon Shelah, undated a. A.E.C. with not too many models. Preprint.

undated b. Compactness in ZFC of the quantifier on "complete embedding of BA's."

undated c. Historic iteration with $\aleph_\varepsilon$-support, *Arch. Math. Logic*, in press.

undated d. Incompactness in singular cardinals.

undated e. Model theory for Θ-complete ultrapowers.

undated f. Non-reflection of the bad set for $I_\theta[\lambda]$ and pcf, *Acta Mathematica Hungarica*, submitted.

undated g. On complicated models. Preprint.

undated h. PCF with narrow choice or ZF + DC + Ax4. Preprint.

undated i. PCF without choice, *Arch. Math. Logic*, submitted.

undated j. Pseudo pcf, *Isr. J. Math.*, submitted.

1975. Categoricity in $\aleph_1$ of sentences in $L_{\omega_1,\omega}(Q)$, *Isr. J. Math.* **20**, 127–148.

1987a. Classification of nonelementary classes. ii. Abstract elementary classes. In J. T. Baldwin, editor, *Classification Theory (Proceedings of the USA–Israel Conference on Classification Theory, Chicago, IL, December 1985)*, volume 1292 of Lecture Notes in Mathematics, pp. 419–497. Springer, Berlin.

1987b. Universal classes. In J. T. Baldwin, editor, *Classification Theory (Proceedings of the USA–Israel Conference on Classification Theory, Chicago, IL, December 1985)*, volume 1292 of Lecture Notes in Mathematics, pp. 264– 418. Springer, Berlin.

1990. *Classification Theory and the Number of Nonisomorphic Models*, volume 92 of Studies in Logic and the Foundations of Mathematics. North-Holland, Amsterdam.

1993a. More on cardinal arithmetic, *Arch. Math. Logic* **32**, 399–428.

1993b. The future of set theory. In *Set Theory of the Reals (Ramat Gan, 1991)*, volume 6 of Israel Math. Conf. Proc., pp. 1–12. Bar-Ilan University, Ramat Gan.

1994a. Bounding $pp(\mu)$ when $cf(\mu) > \mu > \aleph_0$ using ranks and normal ideals. In *Cardinal Arithmetic*, volume 29 of Oxford Logic Guides. Oxford University Press.

1994b. *Cardinal Arithmetic*, volume 29 of Oxford Logic Guides. Oxford University Press.

1996. Further cardinal arithmetic, *Isr. J. Math.* **95**, 61–114.

1997a. The pcf-theorem revisited. In Ronald L. Graham and Jaroslav Nešetřil (editors), *The Mathematics of Paul Erdős, II*, volume 14 of Algorithms and Combinatorics, pp. 420–459. Springer, Berlin.

1997b. Set theory without choice: not everything on cofinality is possible, *Arch. Math. Logic* **36**, 81–125. A special volume dedicated to Professor Azriel Levy.

2000a. Applications of pcf theory, *J. Symbolic Logic* **65**, 1624–1674.

2000b. The generalized continuum hypothesis revisited, *Isr. J. Math.* **116**, 285–321.

2000c. On what I do not understand (and have something to say), model theory, *Math. Japonica* **51**, 329–377.

2000d. On what I do not understand (and have something to say:) part I, *Fundam. Mathematicae* **166**, 1–82.

2000e. Strong covering without squares, *Fundam. Mathematicae* **166**, 87–107.

2002. Pcf and infinite free subsets in an algebra, *Arch. Math. Logic* **41**, 321–359.

2003a. Logical dreams, *Bull. Am. Math. Soc. (N.S.)* **40**(2), 203–228.

2003b. NNR revisited. Preprint.

2006. More on the revised gch and the black box, *Ann. Pure Appl. Logic* **140**, 133–160.

2007. Power set modulo small, the singular of uncountable cofinality, *J. Symbolic Logic* **72**, 226–242.

2008. Theories with Ehrenfeucht-Fraïssé-equivalent non-isomorphic models, *Tbilisi Math. J.* **1**, 133–164.

2009a. *Classification Theory for Abstract Elementary Classes*, volume 18 of Studies in Logic: Mathematical logic and foundations. College Publications, London.

2009b. *Classification Theory for Abstract Elementary Classes 2*, volume 20 of Studies in Logic: Mathematical logic and foundations. College Publications, London.

2009c. Model theory without choice: categoricity, *J. Symbolic Logic* **74**, 361–401.

2010. Large continuum, oracles, *Central Eur. J. Math.* **8**, 213–234.

2011a. The character spectrum of $\beta(n)$, *Topology and its Applications* **158**, 2535–2555.

2011b. Models of expansions of ∞ with no end extensions, *Math. Logic Q.* **57**, 341–365.

2012. Nice infinitary logics, *J. Am. Math. Soc.* **25**, 395–427.

Saharon Shelah and Jindrich Zapletal, 1999. Canonical models for $\aleph_1$ combinatorics, *Ann. Pure Appl. Logic* **98**, 217–259.

Alexandra Shlapentokh, 1989. Extension of Hilbert's tenth problem to some algebraic number fields, *Commun. Pure Appl. Math.* **42**(7), 939–962.

2007. *Hilbert's Tenth Problem. Diophantine Classes and Extensions to Global Fields*, New Mathematical Monographs, volume 7. Cambridge University Press.

2008. Elliptic curves retaining their rank in finite extensions and Hilbert's tenth problem for rings of algebraic numbers, *Trans. Am. Math. Soc.* **360**(7), 3541–3555.

Joseph R. Shoenfield, 1967. *Mathematical Logic*. Addison-Wesley, Reading, MA.

Wilfried Sieg, 2006a. Gödel on computability, *Philos. Math.* **14**, 189–207.

2006b. Step by recursive step: Church's analysis of effective calculability. In *Church's Thesis after 70 years*, volume 1 of Ontos Math. Log., pp. 456–490. Ontos Verlag, Heusenstamm.

Hava T. Siegelmann and Eduardo D. Sontag, 1995. On the computational power of neural nets, *J. Comput. System Sci.* **50**(1), 132–150.

Waclaw Sierpinski, 1956. *L'Hypothèse du Continu*. Chelsea, New York.

Th. Skolem, 1923. Einige Bemerkungen zur axiomatischen Begründung der Mengenlehre. Akateeminen Kirjakauppa. English translation in Van Heijenoort (1967a).

1934. *Ein Verfahren zur Behandlung gewisser exponentialer Gleichungen und diophantischer Gleichungen*, volume 8, pp. 163–188 (in German). 8 Skand. Mat.-Kongr., Stockholm.

Stephen Smale, 1961. Generalized Poincaré's conjecture in dimensions greater than four, *Ann. Math.* **74**(2), 391–406.

Robert I. Soare, 1996. Computability and recursion, *Bull. Symbolic Logic* **2**(3), 284–321.

2004. Computability theory and differential geometry, *Bull. Symbolic Logic* **10**(4), 457–486.

Sashi Mohan Srivastava, 1988. *A Course on Borel Sets*. Springer, Berlin.

Daniel T. Stallworth and Fred W. Roush, 1997. An undecidable property of definite integrals, *Proc. Am. Math. Soc.* **125**(7), 2147–2148.

Richard Stanley, 2010. Decidability of chess on an infinite board, MathOverflow, July 20. Available at: http://mathoverflow.net/questions/27967.

John Steel, 2000. Mathematics needs new axioms, *Bull. Symbolic Logic*, **6**(4), 422–433.

2004. Generic absoluteness and the Continuum Problem, Slides for talk given at the Laguna workshop on philosophy and the Continuum Problem. Available at: http://math.berkeley.edu/steel/talks/laguna.ps.

2009. The derived model theorem. In S. Barry Cooper *et al.* (editors), *Logic Colloquium 2006*, pp. 280–327. Cambridge University Press.

John Steel and W. Hugh Woodin, 2012. HOD as a core model. In A. Kechris, B. Loewe, and J. Steel (editors), *Ordinal Definability and Recursion Theory: the Cabal Seminar*. Cambridge University Press, Cambridge.

D. Stewart, 1814. *Elements of the Philosophy of the Human Mind*, Part I (1792), Part II. William Tegg and Company, London.

M. E. Szabo, 1969. *The Collected Papers of Gerhard Gentzen*. North-Holland, London.

William Tait, 1986. Truth and proof: the Platonism of mathematics, *Synthese* **69**, 341–370.

    1998. Foundations of set theory. In H. G. Dales and Gianluigi Oliveri (editors), *Truth in Mathematics*, pp. 273–290. Clarendon Press, Oxford.

    2005. *The Provenance of Pure Reason*, Logic and Computation in Philosophy. Oxford University Press, New York.

    2008. The five questions. In V. F. Hendricks and H. Leitgeb (editors), *Philosophy of Mathematics: Five Questions*, pp. 249–263. Automatic Press/VIP, Copenhagen.

    2010. Gödel on intuition and on Hilbert's finitism. In S. Feferman, C. Parsons, and S. Simpson (editors), *Kurt Gödel: Essays for His Centennial*, Lecture Notes in Logic, pp. 88–108. Cambridge University Press.

Alfred Tarski, 1951. *A Decision Method for Elementary Algebra and Geometry*, 2nd edition. University of California Press, Berkeley, CA.

    1956. *Logic, Semantics, Metamathematics*, translated by J. H. Woodger. Oxford University Press, New York.

Richard Tieszen, 1998. Gödel's path from the incompleteness theorems (1931) to phenomenology (1961), *Bull. Symbolic Logic* **4**, 181–203; reprinted in Tieszen (2005).

    2002. Gödel and the intuition of concepts, *Synthese* **133**, 363–391; reprinted in Tieszen (2005).

    2005. *Phenomenology, Logic, and the Philosophy of Mathematics*. Cambridge University Press, Cambridge.

A. M. Turing, 1936. On computable numbers, with an application to the Entscheidungsproblem, *Proc. London Math. Soc. 2* **42**, 230–265. Erratum: *Proc. London Math. Soc.* 2, 43 (1937) 544–546.

Jouko Väänänen, 2007. *Dependence Logic*, volume 70 of London Mathematical Society Student Texts. Cambridge University Press, Cambridge.

    2012. Second order logic or set theory? *Bull. Symbolic Logic* **18**(1), 91–121.

Jouko Väänänen and Wilfrid Hodges, 2010. Dependence of variables construed as an atomic formula, *Ann. Pure Appl. Logic* **161**(6), 817–828.

Mark van Atten and Juliette Kennedy, 2003. On the philosophical development of Kurt Gödel, *Bull. Symbolic Logic* **9**(4), 425–476.

    2009. Gödel's modernism: on set-theoretic incompleteness, revisited. In *Logicism, Intuitionism, and Formalism*, volume 341 of Synthese Library, pp. 303–355. Springer, Dordrecht.

J. Van Heijenoort, 1967a. *From Frege to Gödel: A Source Book in Mathematical Logic, 1879–1931*. Harvard University Press, Cambridge, MA.

    1967b. Logic as language and logic as calculus, *Synthese* **17**, 324–330.

O. Veblen, 1903. Hilbert's foundations of geometry, *Monist* **13**, 303–309.

    1904. A system of axioms for geometry, *Bull. Am. Math. Soc.* **5**, 343–384.

F. Viète, 1983. *The Analytic Art: Nine studies in algebra, geometry, and trigonometry from the Opus restitutae mathematicae analyseos, seu, Algebrâ novâ.* Kent State University Press, Kent, OH.

I. A. Volodin, V. E. Kuznecov, and A. T. Fomenko, 1974. The problem of the algorithmic discrimination of the standard three-dimensional sphere, *Usp. Mat. Nauk* **29**(5), 71–168 (in Russian). Appendix by S. P. Novikov.

J. von Neumann, 1925. Eine Axiomatisierung der Mengenlehre. *J. reine angewand. Math.* **154**, 219–240; English translation in Van Heijenoort (1967a).

Maxim Vserminov, editor. Hilbert's tenth problem page. Website created under the supervision of Yuri Matiyasevich, http://logic.pdmi.ras.ru/Hilbert10.

Hao Wang, 1961. Proving theorems by pattern recognition—II, *Bell System Tech. J.* **40**(1), 1–41.

1974. *From Mathematics to Philosophy.* Routledge and Kegan Paul, London.

1977. Large sets. In R. E. Butts and Jaako Hintikka (editors), *Logic, Foundations of Mathematics and Computability Theory (Proc. Fifth Int. Congress on Logic, Methodology and Philosophy of Science, University of Western Ontario, 1975), Part I*, University of Western Ontario Series on the Philosophy of Science, volume 9, pp. 309–333. Reidel, Dordrecht.

1985. Two commandments of analytic empiricism, *J. Philos.* **82**, 449–462.

1988. *Reflections on Kurt Gödel.* MIT Press, Cambridge, MA.

1996. *A Logical Journey: From Gödel to Philosophy.* MIT Press, Cambridge, MA.

Paul S. Wang, 1974. The undecidability of the existence of zeros of real elementary functions, *J. Assoc. Comput. Mach.* **21**, 586–589.

Shmuel Weinberger, 2005. *Computers, Rigidity, and Moduli, M. B. Porter Lectures.* Princeton University Press, Princeton, NJ.

H. Weyl, 1918. *Das Kontinuum.* English translation by Stephen Pollard and Thomas Bole, *The Continuum.* Dover, Mineola, NY, 1994.

A. Whitehead and B. Russell, 1910. *Principia Mathematica*, three volumes. Cambridge University Press, Cambridge.

Mark Wilson, 2008. *Wandering Significance: An Essay on Conceptual Behaviour.* Oxford University Press, New York.

W. Hugh Woodin, 1988. Supercompact cardinals, sets of reals, and weakly homogeneous trees, *Proc. Natl. Acad. Sci. USA* **85**(18), 6587–6591.

2001a. The continuum hypothesis. I, *Notices Am. Math. Soc.* **48**(6), 567–576.

2001b. The continuum hypothesis. II, *Notices Am. Math. Soc.* **48**(7), 681–690.

2002. Correction: "The continuum hypothesis. II", *Notices Am. Math. Soc.* **49**(1), 46.

2010a. *The Axiom of Determinacy, Forcing Axioms, and the Nonstationary Ideal*, revised edition, volume 1 of de Gruyter Series in Logic and its Applications. Walter de Gruyter, Berlin.

2010b. Suitable extender models I, *J. Math. Logic* **10**, 101–341.

2011. The continuum hypothesis, the generic multiverse of sets, and the Ω-conjecture. In J. Kennedy and R. Kossak (editors), *Set Theory, Arithmetic, and Foundations of Mathematics*, ASL Lecture Notes in Logic, pp. 13–42. Cambridge University Press.

J. W. Young, 1911. *Lectures on Fundamental Concepts of Algebra and Geometry.* Macmillan, New York.

Ernst Zermelo, 1913. Über eine Anwendung der Mengenlehre auf die Theorie des Schachspiels, *Proc. Fifth Congress Mathematicians (Cambridge 1912)*, pp. 501–504 (in German); English translation by Ulrich Schwalbe and Paul Walker, Zermelo and the early history of game theory, *Games Econ. Behav.* **34**, 2001, 123–137.

1930. Über Grenzzahlen und Mengenbereiche. Neue Untersuchungen über die Grundlagen der Mengenlehre, *Fundam. Mathematicae* **16**, 29–47.

Index

Abfolge relation, 83–85, 88, 90–93, 98–100
Ableitbarkeit, 81–83, 86, 90–93, 95, 98–100
abstract elementary classes, 123, 250, 252, 254–255
Axiom of Choice, 112, 185–186, 190, 247, 253
Axiom of Constructibility, 112, 125
Axiom of Determinacy, 162, 171–172, 175, 177, 179, 244, 248
axiomatization, 61, 63–72, 76–77, 120, 183, 202, 206

Bagaria, J., 7
Bernays, P., 5–6, 59, 80, 101, 116
Bolzano, B., 81–95, 98–100, 103–105
Boolos, G., 1, 3, 23, 186
Borges, J. L., 78, 106
Brouwer, L. E. J., 15, 39, 47–48, 50–51, 183
Burgess, J., 4, 6–7, 11, 14, 38

Cantor, G., 18, 21, 35
Carnap, R., 24, 131–132, 135–136
Carson, E., 45
Church, A., 112–113, 115–116, 119, 212–213, 215, 257, 261
Church's thesis, 117
Cohen, P., 6, 26, 29, 153, 170, 178, 180, 184, 186, 242
Collatz problem, 235, 241
completeness, 2–3, 59–60, 62, 68–72, 75–77, 80, 94, 97–106, 160–161, 170, 181
completeness theorem, 2, 76, 80, 98, 101, 104, 123, 181, 206, 215, 250
Comte, A., 69–70
constructivism, 47, 50, 53, 147
continuum hypothesis, 6–8, 11, 14–17, 21, 25–26, 28–30, 37, 109, 112, 125, 138, 140, 145, 153, 163, 165, 167, 169–171, 177–178, 184, 186, 188, 194, 196, 201, 211, 244, 246

Davis, M., 110, 115, 211, 227, 256, 259
Dawson, J., 101
Dedekind, R., 49, 144, 147
deductive system, 62

Dehn, M., 219–220
dependence logic, 197–199, 202–203, 256
determinacy, 26, 148, 156, 162–164, 171–173, 176, 179, 243, 253, 256
Dzamonja, M., 5

empiricism, 132
Entscheidungsproblem, 119, 215, 257
Euclidean geometry, 12, 15, 49, 68, 149, 183
extender models, 175–177

Feferman, S., 7, 116, 257
Feng, Q., 172
finitism, 39
forcing, 7–8, 23, 112, 123–124, 155–157, 161–163, 166–168, 171, 173, 175, 197, 207, 242, 245–246, 248, 253
formalism, 15, 104, 110–113, 116–121, 123–124, 126
formalization, 59, 61, 96, 110, 118, 178
Foucault, M., 78–79
Frege, G., 2, 15, 19, 22, 24, 49–53, 80, 103–104
Friedman, H., 7
Friedman, M., 46, 131

Gandy, R., 113, 117–118
generalized continuum hypothesis, 112, 171, 177, 245, 253
generic absoluteness, 8, 123, 160–161, 164, 169, 173–174, 179
generic multiverse, 170, 180, 192, 194, 200, 207, 256
Gentzen, G., 92–100, 105, 257
Gergonne, J, 69–70
Gibbs Lecture, 1, 3, 8, 120–121, 130, 133–135, 137, 141, 144

Hadamard, J., 27
Hahn, H., 15
Hallett, M., 39, 42
halting problem, 110, 213, 215–217, 221, 240–241
Hatami, H., 218

Helmholz, H., 47
Herbrand, J., 80
Heyting, A., 65–68
Higman embedding theorem, 221, 227
Hilbert, D., 8, 15, 39, 61, 72–73, 75, 101–104, 109, 116, 119, 131, 215, 243
Hilbert's program, 158, 256
Hilbert's tenth problem, 211, 213, 218, 227–231, 236, 240, 259–260, 273
HOD, 171
Huntington, E., 63–64, 74
Husserl, E., 21, 33, 47, 136

incompleteness theorem, 3, 24, 76, 114–115, 136, 180, 196, 206, 213
intuitionism, 4, 15, 50, 131, 137

Kant, 1, 12–13, 15, 32–33, 37–40, 42–47, 49–54, 66, 126, 134
Kechris, A., 172
Koellner, P., 7, 131, 148–150
Kreisel, G., 109, 114, 125, 158, 184

Lagrange, J., 85, 90, 100
Laplace, P., 86, 89–90, 100
large cardinals, 7–8, 25–26, 29, 121, 123, 128, 142, 150, 162, 165–167, 169, 174, 177, 186, 195–196
Laver, R., 165–166, 207
Leibniz, 1, 3, 33, 132
Lévy Reflection Principle, III
Lévy, A., 26, III, 162–163, 258
Lindström, P., 187–188, 251, 254
logical consequence, 2, 4, 81–82, 90–98, 100
logicism, 36, 131

Macintyre, A., 229
Maddy, P., 2, 4, 7, 16, 18, 27–29, 146, 256
Magidor, M., 172, 248, 256
Mahlo cardinals, 22, 25, 140, 145
Main Gap theorem, 4, 248, 250, 254
Markov, A., 113, 221–222, 224
Martin, D.A., 4, 7, 16, 19, 24, 26, 140, 144–145, 172
Martin's Axiom, 26, 244
Martin's maximum, 165, 245
Matiyasevich, Y., 243
measurable cardinal, 22, 26, 140, 156, 159–160
Mostowski, A., 112, 186
multiverse, 4, 7–8, 167–169, 177, 179, 182, 185, 194–195, 197, 203
multiverse dependence logic, 198, 202, 206–208
multiverse language, 164, 166–171, 174, 178
multiverse model, 191–192, 194

Neeman, I., 163, 246
nominalism, 7, 14, 104

Novikov, P. S., 220, 222, 224

Parsons, C., 4, 15, 17–18, 20, 22, 24, 30, 32–33, 35–36, 38–39, 46, 110, 112, 124, 131
Peano, G., 16
Perelman, G., 223
Platonism, 1–3, 5–7, 109, 136
Poincaré, H., 36, 39, 47–50, 54, 134
Polya, G., 27
Port Royal Logic, 67, 256
Post correspondence problem, 216, 218, 221
Principia, 59–62, 64–65, 67–68, 76–77, 114, 132–134, 143–144, 256
projective uniformization, 7
Proper Forcing Axiom, 157, 246
Pudlak, P., 116
Putnam, H., 11, 22, 227, 243, 256

Quine, W.V., 131–136

Rautenberg, W., 112
realism, 4, 6, 32–35, 37–39, 41, 44, 46–47, 51–55, 110, 135, 138, 142–143, 146–147, 149, 257
reflection principle, 155
revised generalized continuum hypothesis, 248, 254
Robinson, J., 227–228, 243
Russell, B., 2, 21, 59–65, 76, 132, 134, 257

Sargsyan, G., 176–177
Schanuel's conjecture, 229
semi-axiom, 243–245, 253
Shelah, S., 4, 184, 246, 248–249
Shoenfield, J. R., 112, 186
Sieg, W., 113–116, 118–119
skepticism, 20, 53, 72, 101
Skolem, T., 80, 181, 183, 219, 251
Solovay, R., 22, 26, 112, 162–163
Steel, J., 4, 7, 153, 163, 192, 194, 197, 207
strong absolutist thesis, 168
supercompact cardinal, 156–157, 164, 174–175, 179

Tait, W., 6, 19, 38–39, 45, 147–150
Takeuti, G., 99
Tarski, A., 16, 100, 112, 178, 186, 197, 229, 247
team semantics, 196–198, 256
Turing machine, 113–115, 117–119, 138, 212, 229, 235, 256
Turing, A., 1, III, 113, 117–118, 212–215

universal Turing machine, 234

van Heijenort, J., 80
Vienna Circle, 15, 135
von Neumann, J., 181, 183, 185–186, 199

Wadge, 173, 176
Wang, H., 33, 39, 55, 80, 105, 118, 125, 127–128, 132, 135–136, 138–139, 141–143, 216–217
Wang, P., 233
weak absolutist thesis, 168
weak generalized continuum hypothesis, 245, 253
weak relativist thesis, 167–168, 170, 175
Weierstraß, K., 16
Weyl, H., 48–50

Whitehead group, 244
Whitehead, A. N., 59–63, 76, 258
Wilkie, A., 229
Wittgenstein, L., 105, 257
Woodin cardinal, 123, 156–157, 159, 161–164, 166, 170, 172–177
Woodin, H., 7–8, 26, 123, 156, 161–166, 168–174, 176–177, 192, 207, 244, 246
word problem, 220–222, 224–225, 227, 260

Printed in the United States
By Bookmasters